UPDATE—REPORT ON THE PLANET EARTH

FRANKLIN INSTITUTE BOOKS

Published by The Westminster Press
for The Franklin Institute of Philadelphia

The Amazing Laser, by Ben Bova

Stretching a Point, by Mitch Struble

The Facts of Flight, by Jerry Grey

The Web of Space-Time, by Mitch Struble

Man the Engineer—Nature's Copycat,
by Theodore W. Munch

Sounds and Signals—How We Communicate,
by Charles T. Meadow

Noise, Noise, Noise!, by Jerry Grey

Submerge! The Story of Divers and Their Crafts,
by Anabel Dean

Update—Report on the Planet Earth,
by Joel N. Shurkin

UPDATE—REPORT ON THE PLANET EARTH

by Joel N. Shurkin

A FRANKLIN INSTITUTE BOOK

THE WESTMINSTER PRESS
PHILADELPHIA

BOOK DESIGN BY DOROTHY E. JONES

PUBLISHED BY THE WESTMINSTER PRESS®
PHILADELPHIA, PENNSYLVANIA

PRINTED IN THE UNITED STATES OF AMERICA

Library of Congress Cataloging in Publication Data

Shurkin, Joel N 1938–
Update—report on the planet earth.

"A Franklin Institute book."
Bibliography: p.
Includes index.
SUMMARY: Discusses the cosmic and geophysical forces that created and continue to move and change the earth.
1. Earth sciences—Juvenile literature. [1. Earth sciences] I. Title.
QE29.S48 551 76–14373
ISBN 0–664–32599–8

For my father

Contents

Introduction

In a perfectly average part of the universe is a galaxy called the Milky Way. It gets that name because the stars in it appear so thick that they look like star clouds, milky wisps of stars across the nighttime sky.

The Milky Way galaxy is spiral in form, appearing from the top to be like a whirlpool. From the side it looks flat, with a glowing center where millions of stars are jumbled together, something like the nighttime sky over a city, but infinitely bigger.

The size of it is almost beyond description. If you took a picture of the galaxy and reduced it to the size of this page, then stuck a pin into the picture, the pinhole would represent an area so vast that a spaceship traveling at the speed of light (which Albert Einstein said is impossible) would take seven hundred years to cross from one side of the pinhole to the other.

A German scientist, Hoimar von Ditfurth, has figured out still another way to describe the size. Think of a cloudy day. There is a drop of rain 7/1000th of an inch in diameter (.018 cm) that falls in New York City. The next raindrop hits 50 miles (80.5 km) north, in Connecticut, the next in Massachusetts, the next in Vermont. If you reduced the stars in the galaxy to the size of the raindrops, you would have to travel those distances to go from star to star.

This is an average galaxy!

About a third of the way from the center of this galaxy is an average-size star. It is normal in every way. There are many stars that are brighter and many stars not so bright.

Like many others, this star has a solar system, a family of

planets that circle around it like baby animals circling their mother.

There are nine planets in this particular system. Some are fairly small, like the one closest to this sun. A few are giants, like the fifth one out.

The solar system is quite large. Take the third planet out: If we shrank that planet to about the size of a grapefruit, 4.5 inches (11.4 cm) in diameter, the star it circles would be a mile (1.6 km) away and would be about 46 feet (14 m) wide.

Circling this planet is a moon, which, if shrunk accordingly, would be about 1.5 inches (3.8 cm) in diameter, orbiting about 12.5 feet (3.8 m) away.

The next planet out would be 2.75 inches (7 cm) across, and would never get closer than 1,640 feet (500 m).

The last planet out would be 37 miles (60 km) away, the distance of a strenuous two-day hike.

And the nearest star to that star would be farther away—on this tiny scale—than our earth is to our moon.

That third planet is fairly small.

Its moon is unusually large for a moon, but that is not the only thing odd about this planet. There is life on it!

Many of the planet's most intelligent species think their planet is quite beautiful. And it is.

Seen from space it is mostly blue and white. The blue is caused by the fact that two thirds of the planet is covered with water. The white is clouds, caused by moisture in the atmosphere.

There is also dry land, large blocks of it, called continents. There is vegetation that turns much of the dry land green.

Both the water and the continents teem with life. It looks quiet and peaceful, except for an occasional atmospheric storm. True, sometimes temperatures go to extremes. There are parts where it is too hot, which is not pleasant to life. And there are parts that are brutally cold. And occasionally there is something called an ice age when the cold spreads over much of the continents. That also can make life difficult, although the highest form of life evolved during one ice age.

But the peace is deceptive. For one thing, the planet was born in fire and that fire is still burning beneath the cool crust. Occasionally it bursts out through the tops of mountains or cracks in the crust. Occasionally the crust shakes violently and whole cities crumble.

Great blocks of land move around, and through the billions of years of its history the face of the planet has always been changing.

That peaceful planet, born in chaos and inferno, rumbles and groans, explodes and shakes.

This is a report on that planet.

The planet is called Earth.

1
In the Beginning

> And the earth was without form, and void;
> and darkness was upon the face of the deep.
> The Book of Genesis

It is black.

It is not a flat color, but a deep, almost three-dimensional black, like velvet or rich felt.

Occasionally there is a point of light, an intense, burning light. A star.

In some places there are clusters called galaxies, huge streams of stars that stretch in distances beyond all describing.

Sometimes the black is broken up by the blue and red glowing clouds called nebulas. They are made up of all kinds of things, mainly gas and dust. The chief element is hydrogen.

Our story begins with one such nebula, five or six billion years ago.

This nebula was round and generally not very dense. But each little particle in the cloud had a little bit of gravity and that made it attractive to particles nearby.

Here and there an atom in the cloud would be drawn toward another atom. Soon more atoms were clustering toward each other. Gradually they all began to be drawn toward the middle. The cloud began to shrink and as it shrank it began to spin.

The spinning got faster and faster and the cloud flattened,

looking now like a giant phonograph record on some cosmic turntable.

Soon the centrifugal force ripped a section of the cloud away. Most of the dust and gas remained with the center but some of it was now circling the main cloud. Smaller and lighter than it was, the main cloud was no longer spinning nearly so madly.

The central cloud continued to contract. Pressures began to build up in its center, and interior temperatures began to rise.

Then it happened. The hydrogen atoms, squeezed by the pressure, excited by the heat, began to fuse atomically with each other, forming helium. The energy released by this fusion lighted the cloud with fire. A star was born.

The fire from this thermonuclear explosion actually blew itself out. The force of heat was stronger than the gravitational pull of the particles, and now the star was expanding rapidly instead of contracting.

But as it expanded the pressures dropped. The temperatures fell. The reaction could no longer sustain itself, so the light went out.

The star, now cold, went through the same contraction process as before, which led again to a thermonuclear explosion and expansion.

We don't know how many times the star went through this cycle. We do know that all young stars seem to grow up that way. And we also know that sometime about four billion years ago, this adolescent period ended and the star stabilized at about its present size. It will probably stay that way for another four billion years. It is our sun.

While this was going on, a similar process was changing that cloud which had been circling the pulsating star. Particles within it had formed at least nine whirlpools orbiting the center, with small eddies within the whirlpools.

As these swirls swept around the sun they gathered more of the gas and dust of the cloud, like a magnet sweeping up iron filings. They grew heavier and thicker, and finally condensed into the planets as we know them.

The tinier eddies within the swirls formed moons. Some planets, like Mercury and Venus, the first and second out from

ASA

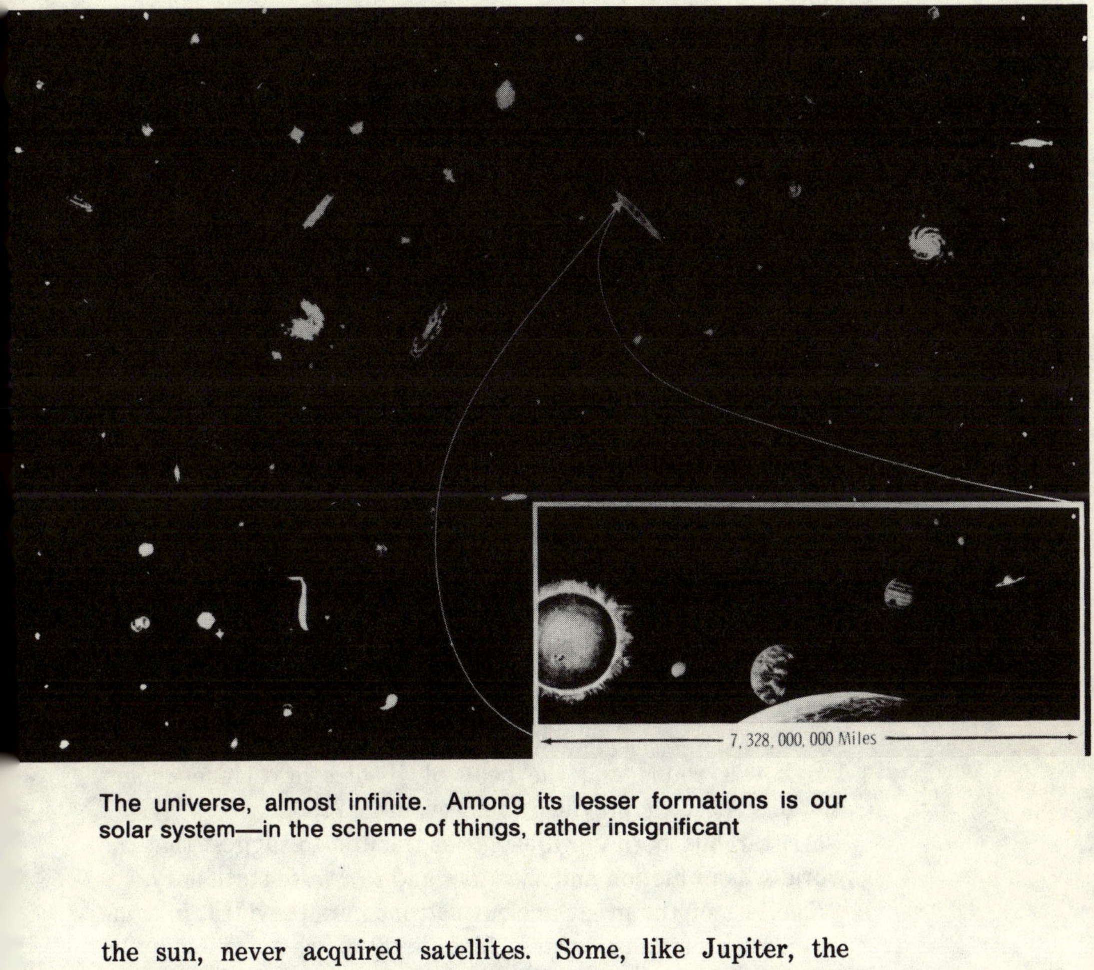

The universe, almost infinite. Among its lesser formations is our solar system—in the scheme of things, rather insignificant

the sun, never acquired satellites. Some, like Jupiter, the largest, found themselves with many.

The planets were formed in somewhat the same way the star was created, each condensing from a cloud. But they never flared up the way the star did, because they lacked the mass to create the kind of pressure and temperature required for thermonuclear reactions.

Jupiter, the fifth planet out, came closest to becoming a sun. Some scientists, in fact, call Jupiter a "failed star." But as big as it is—it contains more mass than all the other planets combined—the planet Jupiter was still not big enough to trigger a solar reaction.

Earth is the third planet out, and one of the smallest. We are fairly sure it was formed about 4.5 billion years ago, sometime after the sun.

So that is the story of how the Earth was formed. Actually, to be more accurate, that is the story as scientists now generally accept it to be. There have been, you see, many theories about the Earth, all described by honest men based on the best information at hand. Most of the scholars were quite specific and quite sure of themselves. As you will see in reading this book scientists are frequently wrong. Indeed, sometimes their degree of inaccuracy is directly proportional to their confidence in what they say.

The origin and nature of our planet has been a matter of conjecture and conflict since men first began to think.

Early man thought the Earth was flat. It looked flat, didn't it? He could feel no motion, so it must stand still. The objects in the sky seemed to have motion. So it was logical to assume we are stationary and the heavens move around us.

Not every ancient accepted this obvious truth. Anaximander, who lived about 570 B.C., was the first we know of to declare that the Earth was suspended in space. Pythagoras declared the Earth was round and the heavenly bodies have independent motion. Aristotle provided some scientific proof.

Aristarchus, born in 310 B.C., was the first to suggest that the world was in motion and the stars and sun were stationary.

The last of the great ancient astronomers provided, in some ways, a long step backward. Claudius Ptolemaeus (Ptolemy), a Greek who lived in Alexandria in the second century A.D., declared the Earth was in fact the center of the universe and everything flew around it. That became the "official word" in the medieval Church, and to contradict Ptolemy was to be a heretic. A student could get into a lot of trouble that way.

The Earth and man, the Church asserted, was at the center of all things—quite an arrogant notion when you think about it.

The first man we know of to publicly but quietly hint that Ptolemy was wrong was Nicholas of Cusa (1401–1464). He suggested that perhaps the Earth was not the center of the

'ASA

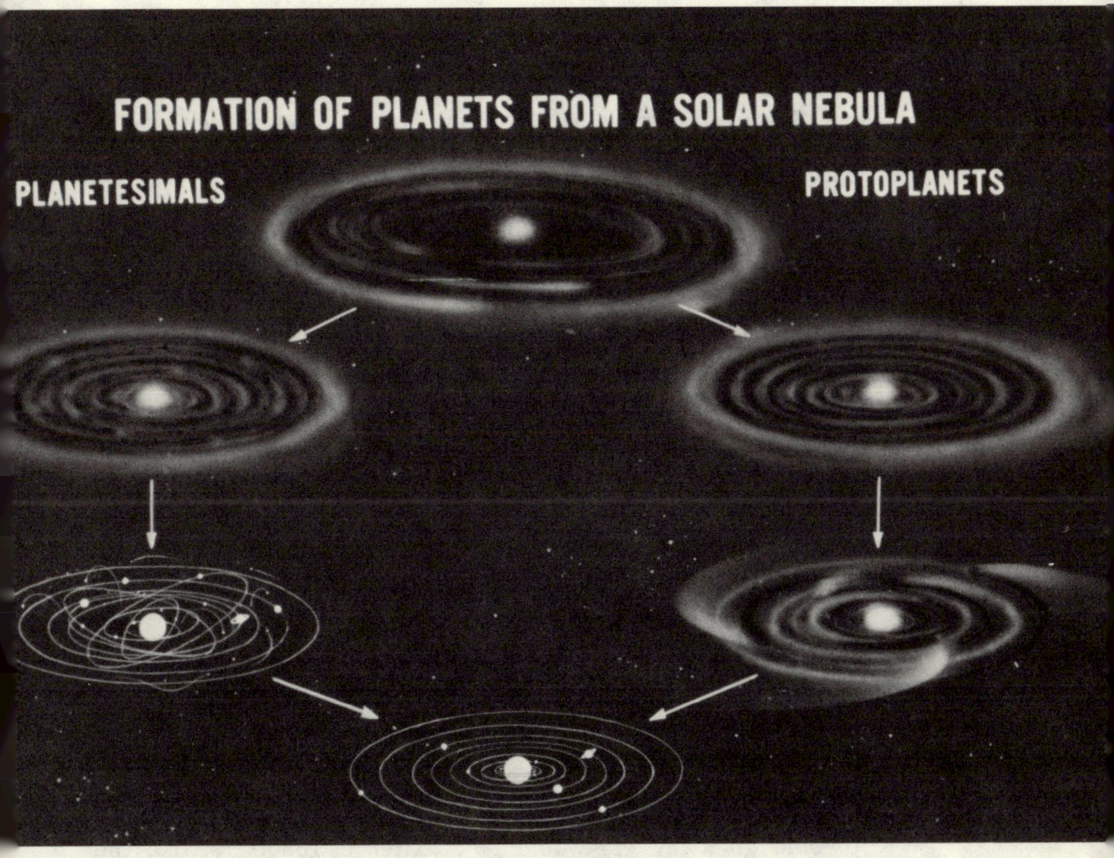

From a gaseous cloud came the solar systems, forming protoplanets (infant worlds) and planetesimals (the junk that litters the rest of planetary space, like comets). This probably happened in more than one place

universe at all. No one paid any attention to him.

Then came Mikolaj Kopernik, or Copernicus, one of those unquestioned giants of humanity.

In 1530 or 1533 Copernicus finished a book called *De Revolutionibus Orbium Coelestium*—"On the Revolutions of Heavenly Bodies." It was not published until 1543 just before his death, when he was relatively safe from heretic chasers.

But even before his death the word had got out that he was

going to contradict Ptolemy. A small volume of his, *Narratio prima*—"First Account"—was dedicated to Pope Paul III and he wrote in it:

"I am fully aware, Holy Father, that as soon as they hear in these volumes of mine about the revolutions of the spheres of the Universe and that I attribute some sort of motion to the Earth, some persons will immediately raise a cry of condemnation against me and my theories."

He was absolutely right. That is exactly what happened, with Martin Luther leading the charge. Luther wrote:

"Mention has been made of some new astrologer, who wants to teach that the Earth moves around, not the firmament or heavens, the Sun and Moon. Just as though a man were to sit in a cart or a moving boat, and thinks he sits motionless and rests while the earth and the trees move. . . . This fool seeks to overturn the whole art of astronomy. But as the Holy Scriptures show, Jehovah ordered the Sun, not the Earth, to stand still."

When *De Revolutionibus* was published it included this famous phrase: "I also say that the Sun remains forever immobile, and that whatever apparent movement belongs to it can be verified as due to the mobility of the Earth."

Or, in other words, we are *not* the center of all things.

The Church tried to ignore Copernicus, hoping his theory would go away. It would not. Later astronomers, like the colorful Tycho Brahe of Denmark who wore a gold and silver nose after losing his real one in a duel, improved upon the theory. And eventually Isaac Newton put the icing on the cake.

The date of creation has been another subject of dispute. Take, for example, the splendid case of James Ussher, the Archbishop of Armagh, a noted scholar of his day.

In 1654 he concluded after deep study that the world was created on Saturday, October 23, 4004 B.C.

At 9 A.M.!

We do not know whether that was Standard Time or Daylight Saving Time.

Because he was a man of authority, his word was law. For more than a hundred years it was considered heresy to contradict the honorable archbishop.

How can we be so sure he is wrong?

The oldest complete fossil record we have goes back about 600 million years. Those fossils are of a time when the main divisions of the plant and animal world had already been created by evolution. The fossils, of course, are much older than Ussher's theory would permit.

We have even older fossils of single-celled creatures like algae. Paleontologists, the name for scientists who study fossils, believe life began on Earth about three billion years ago.

We have found Earth rocks almost four billion years old, a date arrived at by studying the radioactivity in them. We know the rate of decay of radioactive elements, and by measuring the amount of radioactivity in the rocks we know how much has decayed, and therefore how old the rocks are.

We know the Earth was molten before those rocks were formed, and so the planet is even older than that.

So we estimate the birthday of the planet to be about 4.5 billion years ago, give or take a couple of million. (You will notice that modern scientists lack the precision—or arrogance—of James Ussher.)

The theory we have described is relatively new. Before it was devised, scientists had other thoughts about how the world was born.

The French astronomer Pierre Simon de Laplace came pretty close in the eighteenth century. His theory, which was also arrived at independently by the German philosopher Immanuel Kant, described planets forming out of clouds circling the sun. The clouds shrank because of gravity and threw off rings which became the planets.

That sounds pretty good, except that Laplace supported the theory with mathematics that later proved incorrect. It was a good try, though.

Laplace was replaced with tidal theories involving the idea that something had drawn matter from the sun.

George Louis Leclerc, in 1745, suggested that the sun had been hit by a passing body. He didn't say what.

Sir James Jeans, in 1901, suggested that a passing star had

sucked matter from the sun and the amputated gases had formed the planets. His math was off too.

The German astronomer C. von Weizsäcker came up with the accretion method of creation in which the sun accumulated matter from a nebula as it passed through. There are still supporters for his theory.

About 1950 Harold Urey, an American, working on ideas suggested by others, came out with the so-called "Cold Earth" theory, which we used to begin this chapter. That theory is the one now generally accepted in some form or other. Urey's theory avoids the mathematical pitfalls of Kant and Laplace.

Where did the moon come from? There were several theories, which can be called the "mother-daughter" theory, the "brother" theory, and the "wandering stranger" concept.

The "mother-daughter" theory originates with Sir George Darwin, the son of Sir Charles, the evolution man. He concluded the moon broke out of the Earth and went into orbit. Indeed, he and many other scientists believed the Pacific Ocean was the crater remaining after the moon burst into space.

The "wandering stranger" theory, which helps account for the basic chemical differences between the Earth and the moon, suggests that the moon was created elsewhere and was wandering by one day when it got caught in the gravitational pull of our planet. Because of some basic laws of physics, the moon is doomed to circle the larger planet, slowly, ever so slowly, being drawn down toward it.

Some scientists like Urey suggested the "brother" theory. This holds that both bodies were formed at the same time. The chemicals on the moon are found in different proportions than they are on the Earth, however. This is because they solidified at different temperatures.

The oldest moon rock brought back by astronauts is 4.25 billion years old (Apollo 16). The moon does not have the weather or the geologic history of the Earth, so rocks stay intact on the moon much longer than do terrestrial rocks.

The problem with explaining the moon is that it is too big. It is not really the typical moon, like the two rough hunks of rock that circle Mars.

Earth, photographed on February 12, 1975, by NASA's Synchronous Meteorological Satellite (SMS), 22,300 miles (35,680 km) up. All of the southwestern United States, Mexico, Baja California, and southern Florida, as well as Central America and Cuba, are readily visible. The sweep of clouds across Florida is a thunderstorm

Scientists now tend to consider the Earth-moon system to be a double planet system, probably formed at the same time and locked eternally in an astronomical embrace.

If the current theories are correct, what happened to our solar system is probably not all that unusual in the universe. We know now that it may be quite common.

We know, for instance, that there are many double and triple stars in the universe. This could mean that the clouds that condensed around those stars split more evenly than did our nebula, or that they were just bigger.

Instead of forming planets there was enough broken-off material to form second and third stars. Also, if the theory is correct, there should be other planets around other stars. (And if there are other planets should there not be life on some of them?)

We believe now there are other planets (and probably other life).

The planets were first discovered by scientists at Swarthmore College in Swarthmore, Pennsylvania, using a fairly small telescope.

The Swarthmore technique was to take thousands of pictures of individual stars and study the way they moved through space by matching the pictures. There are ways of detecting an unknown object in space by the way a known object moves—you look for a wobble. In fact, the existence of the planet Pluto was first suggested because of the way other planets around our sun went about their orbits.

In 1942, K. A. Strand at Swarthmore found there was probably a third object going around the double (or binary) star 61 Cygni. The object would be too small to be a star, so Strand said it had to be a planet.

Peter van de Kamp, also at Swarthmore, later found two unseen companions to Barnard's star, presumably planets. We think now there are other planets around the stars with the far-out names Lalande 21185A, or AOe 17415-6.

So the theory looks good, according to the evidence. We think we can say that the Earth was formed out of a condensing cloud 4.5 billion years ago.

We think that it eventually became molten, probably because of the radioactivity and gravity of the material that formed it.

The Earth, then, was a ball of fire whirling around the young sun.

Processes had already begun to turn it into the planet that we know.

2
Fire and Rain

> The sky is changed!—and such a change! Oh night,
> And storm, and darkness, ye are wondrous strong,
> . . . Far along,
> From peak to peak, the rattling crags among
> Leaps the live thunder!
>
> Lord Byron

So now we have a great ball of fire.

The Earth is molten, heated by the radioactivity of its elements and gravity pressure.

Before we go on, let's talk about time and the way some scientists think about it.

We speak in terms of days, weeks, months, and years. In talking about history we sometimes refer to centuries.

But to a geologist and a geophysicist centuries are not worth talking about. They speak in terms of millions of years, hundreds of millions of years, and even billions of years.

If we are to chronicle the youth of Earth, we must adopt their habit of gulping down enormous hunks of time as if they were trivial. We must remember that, to a planet 4.5 billion years old, the entire life-span of the human race is a blink of the cosmic eye. We hardly count at all.

The processes that made the world what it is today involve cooling the molten planet down to a seemingly solid ball and forming layers of various kinds of materials. The Earth is, in some ways, like an onion, with layers over layers over layers. It got that way because of the cooling process.

Like everything else about the very early days of Earth, what really happened is a mystery. There are only educated guesses, some of which contradict each other.

For instance, we described the Earth as a ball of fire. Well, some scientists do not think it was ever completely molten, although all seem to agree most of it was at one time.

The cause of the heating was mainly radioactivity but some was the effect of gravity, which pulled material together under great pressure.

Gravity alone could have brought the temperature at the center of the Earth up to 1,800° F (1,000° C). Radioactivity drove it much higher than that.

We know that the core of the Earth, which is mostly iron and nickel, was molten and it may have taken between 500 million and one billion years for temperatures to get high enough to melt that material.

Whatever it was like, all melted or only partially melted, Earth began to cool down and form layers.

Certain elements began moving around as they cooled. The iron and nickel sank toward the center. Lighter elements, like silica, floated up toward the top, a process that took between 4.5 and 5 million years or so.

The first thing that happened is that two interior sections, which we call the core and the mantle, separated. This is called the Iron Catastrophe.

In the 1930's a Dutch scientist, Vening Meinesz, offered one suggestion about how this separation took place.

Meinesz did not believe the surface of Earth was ever completely molten. He thought some of it had already hardened when the core was formed.

The Iron Catastrophe took place because of something called convection currents, he said. Convection currents, incidentally, an important concept to remember, are acting when water boils in a pot. The surface of the water bubbles because heat is rising in currents—convection currents—bringing water from the bottom to the top.

In geophysics, convection currents refer to currents of heat rising through molten material, carrying material with them.

Meinesz said that during Earth's early years, there was no core, only a single system of convection currents. The heavy stuff sank toward the center, pushed by the currents, and the lighter stuff became the "froth" on top, forming the first continents as it hardened, like fat coming to the surface of cooling gravy.

As the heavy material formed a core at the center, the convection currents had less room to move around and the flow of heat toward the surface was broken up. This moved the crust, breaking up the land mass and creating gaps between them, something like oceans, but without water.

For about a billion years after the core and mantle separated, the crust was solidifying, melting, and solidifying again, scientists now believe.

Part of the chaos was caused by volcanoes, which burst through the crust and shot the hot, molten material all over the rocks.

Now think of seas of glowing red-orange lava, bubbling, gurgling, and streaming around black and brown rocks. The smell must have been terriffic, with fumes much like what we think of when we think of hell.

There was smoke and there was fire and there was brimstone. Earthquakes racked the land.

Gases steamed out of the Earth to form the first atmosphere. It was an odoriferous concoction of methane (natural gas), hydrogen, and water vapor (steam).

There were probably titanic storms and eventually boiling rain fell on the new rocks as soon as temperatures cooled down enough to permit condensation.

It must have been an awful place.

About 3.9 billion years ago, just yesterday, the crust calmed down some. The first continental rocks, granite, were formed and we have actually found some of them.

This doesn't mean the storm and fire ended. They haven't ended entirely to this day. Certainly when the continental granite was finally formed huge volcanic eruptions and melting were still going on as convection currents poured heat from the

Geological Survey

The early Earth must have looked like this, bubbling, exploding molten rock. This photograph was made in the heart of a Hawaiian volcano at night

interior outward.

And there was—and is—plenty of heat.

To be molten, completely or partially, we are talking about some very high temperatures.

At the place where the mantle meets the outside of the core, 1,800 miles (2,900 km) down, temperatures range from about 7,200° to 9,000° F (4,000°–5,000° C). That is hot enough to keep the nickel-iron in the core melted and is about the point where silicates begin to solidify.

It was probably hotter than that when the surface was molten.

The rain that fell on the cooling surface gathered in the depressions between continents, forming the first oceans, which were probably not salty at first.

The rain that fell on the rocks found natural paths to the low places, creating the first rivers and streams.

The continents are probably resting on mantle material. One scientist, Harry Hess of Princeton, believed that the ocean beds are not made of continental rocks at all, but of mantle material, which comprises 80 percent of the Earth's volume.

Now we have a cooling Earth, with a solid crust, a less solid mantle, and a molten core surrounded by a nut of very solid stuff. It is a planet different from some in the solar system.

Let's take a look at the Earth as we know it now. Very likely it has not changed much from those early days in respect to size, shape, and its place in the scheme of things.

For one thing, the Earth is not entirely round. It bulges a bit around the middle and at the top.

It is the third planet out from the sun, after Mercury and Venus, but on the sunward side of Mars. Its maximum distance from the sun is 94,537,000 miles (152,110,000 km). The closest it gets in its nearly circular orbit is 91,377,000 miles (147,026,000 km). The planet travels at a speed of 18.5 miles (30 km) a second or 66,000 mph (106,194 kph). The diameter at the equator is 7,927 miles (12,755 km) and the diameter at the poles is 7,900 miles (12,711 km)—which shows it really does have a potbelly.

Earth travels around the sun in 365.24 days, and every day on Earth is now 24 hours, 3 minutes, and 56 seconds. We said that is the length of the day *now* because the days on Earth are getting longer. This is due to tidal friction, the friction caused by the oceans rubbing against the ocean floors.

This friction is slowing down the rotation of the Earth. It is not exactly noticeable. Every day on Earth is 0.00000002 of a second longer than the day before.

The tidal friction is also causing the moon to move toward the Earth by about 4 inches (10 cm) a month.

Do you want to know how much the Earth weighs? How about 6,600 million million million tons (or 6,600,000,000,000,000,000,000 in numbers)?

That makes the Earth a fairly small planet as the sun's children go. Venus, Mercury, and Mars are smaller but all of the rest of the planets (except maybe Pluto, the outermost) are bigger, some by a great deal.

What is special about Earth and some of its neighbors is that it is fairly solid, with a rocky surface. We call planets like that terrestrial. All the inner planets are terrestrial—Mercury, Venus, Earth, and Mars. Most of the outer planets—Jupiter, Saturn, Uranus, and Neptune, are called Jovian, because, like Jupiter, they are mostly, or entirely gaseous—balls of gas. They have little or no rocks or surface as such.

We do not know enough about Pluto to say, but it appears Pluto may be terrestrial.

With the possible exception of Mars, none of the planets, terrestrial or Jovian, are very hospitable to life. Mars may have life of a primitive sort (we are beginning to send robot spacecraft to find out), or it may have had more advanced life long ago, but it does not have it now.

Mars has probably had water at one time or another in its past and may have had an atmosphere richer than the thin wisp that now makes up Martian air. Mars lost its atmosphere because it is not big enough to hold gas molecules in its gravitational grasp.

Mars is colder than Earth because it is farther from the sun, but noontime on the Martian equator reaches a balmy 70° F (21° C). On the other hand, it can get more than a little chilly, with minimum temperatures down to −94° F (−70° C).

Little Mercury is a horrible place, so close to the sun that it is broiled. Temperatures on Mercury have been measured at up to about 770° F (410° C), which is too hot for any life we know of.

Venus is just as bad. Besides having an atmosphere of carbon dioxide, it gets temperatures at the surface of up to 400° F (204° C).

The gas giants like Jupiter are completely uninhabitable by terrestrial life, with atmospheres of ammonia and methane. Can you imagine what that must smell like?

So our little planet is really quite comfortable.

From that flaming ball it cooled down to a world of temperate climes, huge oceans, and continents teeming with life found nowhere else in the solar system.

A very nice place.

It is time to take a trip through it.

3
A Journey to the Center of the Earth, More or Less

> **Neither you nor anyone else knows for certain what goes on in the interior of the globe, considering that we are familiar with scarcely a 12/1000th part of its radius.**
>
> **Jules Verne, *Journey to the Center of the Earth***

Professor Hardwigg, the adventurous geologist in Verne's delightful science fiction tale, was absolutely right. We don't know for sure what "goes on in the interior of the globe."

We can guess.

We can guess because the Earth is a living thing. It moves and crashes about. Every time it moves, as in an earthquake, it sends out waves which scientists can pick up on delicate instruments.

These waves act in peculiar ways when they run into certain kinds of material. So by monitoring the behavior of the waves scientists can get a pretty good notion of what lies below the crust of the Earth.

Sometimes great tragedies can be scientifically productive. On May 21 and 22, 1960, a colossal earthquake struck Chile. The planet literally rang like a gong. Scientists learned a great deal about the interior of the planet by tracing the waves of that shock.

The hero in Verne's book had to literally climb to the interior of the Earth to find out what is there. He crawled down into an

Iceland volcano.

We do not have to do anything that extreme.

Let's pretend, though. Let us suppose we had the power to drill a hole from where we are standing, straight down into the center of the Earth.

What we are standing on is called the crust. It is not a very significant part of the Earth, representing only about 0.4 percent of the mass of the planet. "Mass" is a scientific word that is complicated to define. If you substitute "weight" for "mass" you will be close to the idea. The rest of the world, all 99.6 percent of it, is generally out of reach to us.

Take the highest mountain on dry land, Everest, in The Himalayas. It is 29,028 feet (8,825 m) above sea level. That's about 5.5 miles (8.8 km).

The crust is between 6 and 25 miles (10–40 km) thick, so Everest is not as high as the crust is thick at its thinnest place.

Spread the crust over the top of the whole Earth and you see that Everest is big only when viewed by tiny creatures like us. It is a mere pimple on the face of our globe.

The continents are made up of granite rocks. The rocks under the ocean floor, if they are crust, not mantle, are generally basalts—fine-grained rocks formed by melting.

We have probed the crust fairly thoroughly and know most of what there is to know about it. It is when you go below the crust that things get a little hazy. But thanks to those seismic waves, we have a general idea.

In 1906, following an earthquake, a geologist from the land that is now Yugoslavia found some weird seismic readings.

His name was Andrija Mohorovicic, and he surmised from his readings that about 20 miles (32 km) below the surface there was an abrupt change in the composition of the Earth. It was thought that this was the boundary between the crust and what lies below, the mantle.

This zone was called the Mohorovicic discontinuity, or Moho to its friends.

The Moho was much closer to the surface under the ocean floor, about 3 to 6 miles (5–10 km), he found, and was very uneven.

It was felt that the mantle was made up of magnesium-iron silicates, those elements that bubbled toward the surface as the Earth cooled during the Iron Catastrophe.

We have found that the top of the mantle includes minerals like olivine and pyroxene, the common ingredients of basic basaltic rocks.

As we drill deeper and deeper, pressures are growing and we get some very rare rocks, like garnets and diamonds. Diamonds are formed only under this kind of intense pressure.

It is also getting very warm. The deepest mines in the world indicate that the temperatures rise about 19° Celsius every kilometer you go down (or 30° F every 0.6 mi.).

The rocks of the mantle below the first layer are the magnesium-iron silicates, which change chemistry somewhat as they get deeper and as pressures grow. The atoms get so squashed together they form crystals that can bend like soft gold.

The mantle is now divided into several different sectors, discernible by the way seismic waves react to them.

The top of the mantle and the crust are fairly rigid, to a depth of about 60 miles (96 km). The American geologist Joseph Barrell called the rigid sector the lithosphere, from the Greek word for stone.

Below that, Barrell said, is a plastic layer (plastic meaning that it bends and shapes) which he called the asthenosphere, from the Greek word for weak. The plates making up the lithosphere ride on top of the asthenosphere.

The asthenosphere begins at about 37 miles (60 km) down and goes to about 155 miles (250 km). It is formed of crystalline rock, the silicates, which has been called a "crystal slush." It is believed that the lava that flows out of volcanoes is from this layer. Most deep earthquakes that are associated with volcanic activity are centered in this zone.

Then begins a zone known as the rocky mantle, which extends about 1,650 miles (2,650 km).

The rocky mantle is a fairly complex structure. If the geologists are right about the many convection currents running through it during the Earth's cooling-down period, then the rocks in the mantle are probably patterned and swirly.

Scientists have found rocks in western Greenland that are almost four billion years old and are unusually complex. These rocks show that the crust had apparently shrunk and the cause of this shrinking was a loss of volume in the mantle, probably because of chemical changes.

It is hard to overemphasize how important the mantle is to the story of the Earth. This red-hot hollow ball of rock makes up almost half of the Earth's radius and more than 80 percent of its volume.

The crust formed from the mantle, and lava from the mantle poured out onto the surface and made new rock. The water in the oceans, lakes, and rivers came in the form of vapor from the mantle. The gases that make up our atmosphere came from the same place.

Life evolved from compounds of the chemical carbon and the mantle was the source of this carbon. Mountains, as we shall see, come from what happens in the mantle. We shall also see how the texture of the mantle affects what happens to continents and leads to earthquakes.

Since we can't go down and visit the mantle, how do we know what it is made of? Believe it or not, one way is to use the indirect method of studying meteorites.

Meteorites, flying hunks of rock in space, were formed at the same time as the solar system and from the same collapsing cloud.

We know about the chemical makeup of the solar system because, using a device called the spectroscope, we know the makeup of the sun. The sun constitutes 99.6 percent of the mass of the solar system, which means that if something is found in the sun it is likely to be found among the planets and the meteorites. If it isn't in the sun, it isn't in its satellites.

There are two kinds of meteorites, those made mostly of iron-nickel and those made of silicates and other forms of rocks.

The kinds and amounts of certain elements in these rocky meteorites are about the same as we expect to find in other bodies in space. So we can compare this to what we know about our planet. We can assume that what is true of the meteorites is

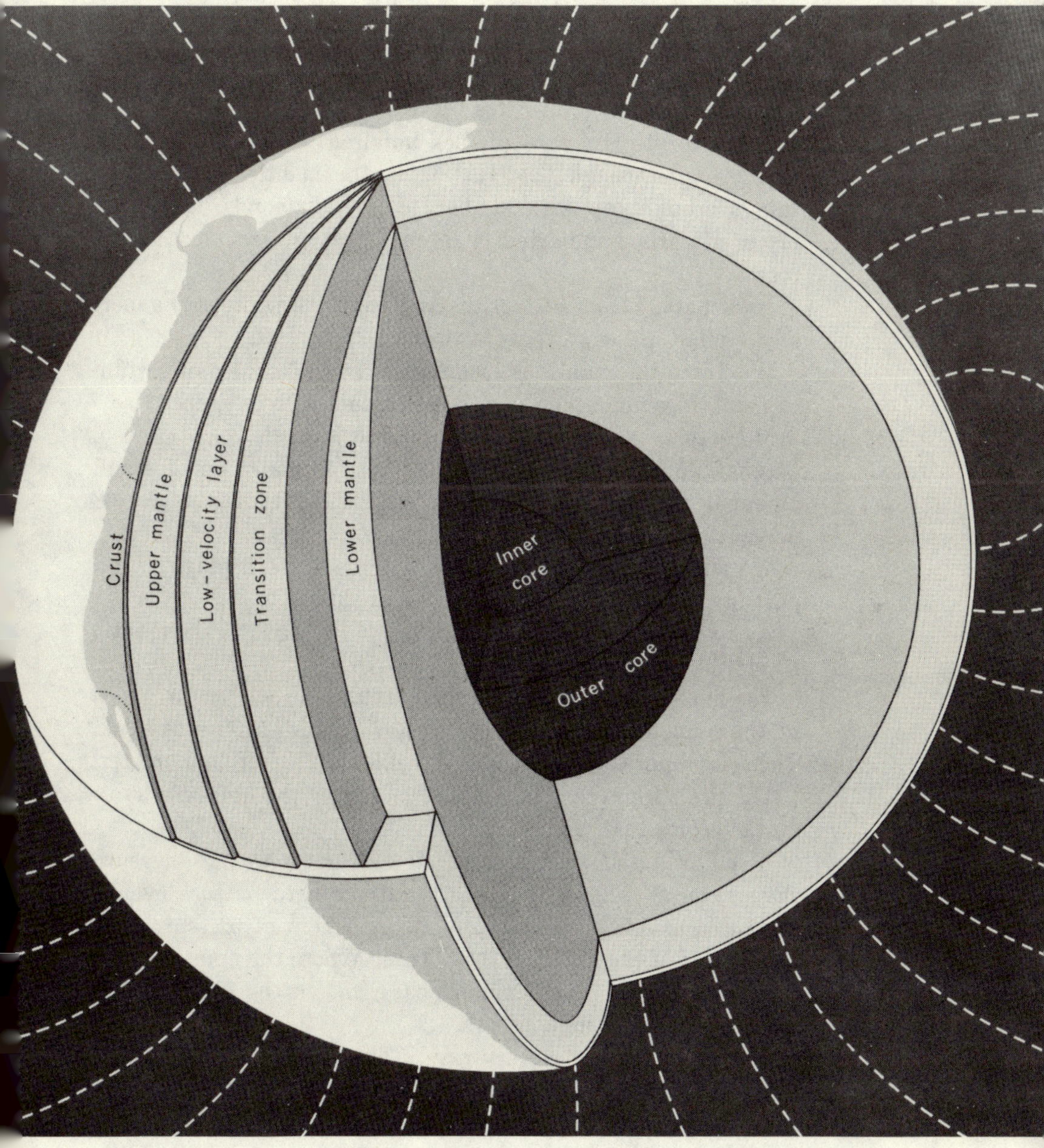

If you took a journey to the center of the world, your road map would look something like this, with the crust, the various layers of the mantle, and finally the heart of Earth, the core, liquid on the outside, solid on the inside, and hot all over

more or less true of the Earth and its main section, the mantle.

We also have some rocks from the mantle available to help in this study. They are nodules or lumps of mantle rock found in formations called kimberlites.

A kimberlite is a tube of rock hundreds of yards thick that punches up through the crust from the mantle. Among the things brought up with kimberlites are diamonds, which are pieces of carbon subjected to intense heat and pressure in the mantle.

Information from meteorites and from the nodules is generally similar, although there are some puzzles left.

We know the mantle is made up mostly of oxides of certain elements—combinations of those elements with oxygen.

More than 90 percent of the mantle is made up of silicon, magnesium, and iron oxides. The rest is made up of small amounts of aluminum, calcium, and sodium oxides. So 98 percent or so is made up of these oxides.

We will get back to the mantle later because it plays a vital part in the great detective story we are about to tell.

At about 1,800 miles (2,900 km) below the surface, the mantle meets the liquid outer core of the Earth.

The liquid core is actually a shell around the solid central core of the planet. The liquid core is 1,375 miles (2,200 km) thick. The solid inner core below it is 760 miles (1,220 km) in diameter.

The pressure here is incredible, caused by the pull of gravity on the atoms that make up the Earth.

K. E. Bullen of Australia has figured out a way to measure the pressure. The weight of the atmosphere at sea level is measured at one bar (a bar is a measurement of pressure). At the bottom of the Pacific it is 800 bars. Where the mantle meets the outer core it is 1.75 million bars, and in the inner core it would be nearly 4 million bars.

The lower mantle is very rigid. The outer core is liquid, a fairly sudden transformation probably caused by the temperature. As we have seen, temperatures there are warm enough to melt the nickel-iron of the core, but not hot enough to melt the silicates.

We think temperatures at the center of the Earth run

between 3,600° and 11,700° F (2,000°–6,000° C).

Again, what we know about the core and mantle is based on seismic soundings, usually earthquakes, but an occasional atomic test has contributed to our knowledge.

In 1974 some scientists at Cornell University discovered something quite unique, a rock that probably came from that molten outer core. The rock is the mineral josephinite, an iron-nickel alloy. It was found in Oregon, just north of the California line.

Inside the rock was a garnet.

Dr. John Bird of Cornell reported that the rock was believed to have been formed 140 million years ago in the Pacific Ocean area. A rare messenger from below.

Our trip is not so exciting as Verne's but it is probably more accurate.

We now go on to something really weird—the case of the reversing footsteps.

4
Dynamo!

Let the great world spin for ever down the ringing grooves of change.

Tennyson

The waves boil and roar. The ship tosses as if a giant hand were lifting it from beneath and slapping it away. Up and over, again and again.

On the deck of the ship the captain, clutching the wooden railing, stared at the ring of numbers in the compass that showed him the direction the storm was chasing his ship. He wanted to head west and, making allowances for the white, pushing storm, he knew he could rely on his compass to show him the way.

The compass upon which the sea captain had staked his life, and the life of his ship and crew, was based on the fact that a magnetized needle will always point north if given freedom of movement. It doesn't point exactly north, but to a point called the magnetic north pole, which is close enough for navigational purposes.

For centuries mariners had relied upon the compass. Until the Elizabethan age it was assumed that the North Star (Polaris to astronomers) or the Great Bear constellation was the attraction—that somehow stars pulled the needle toward the north.

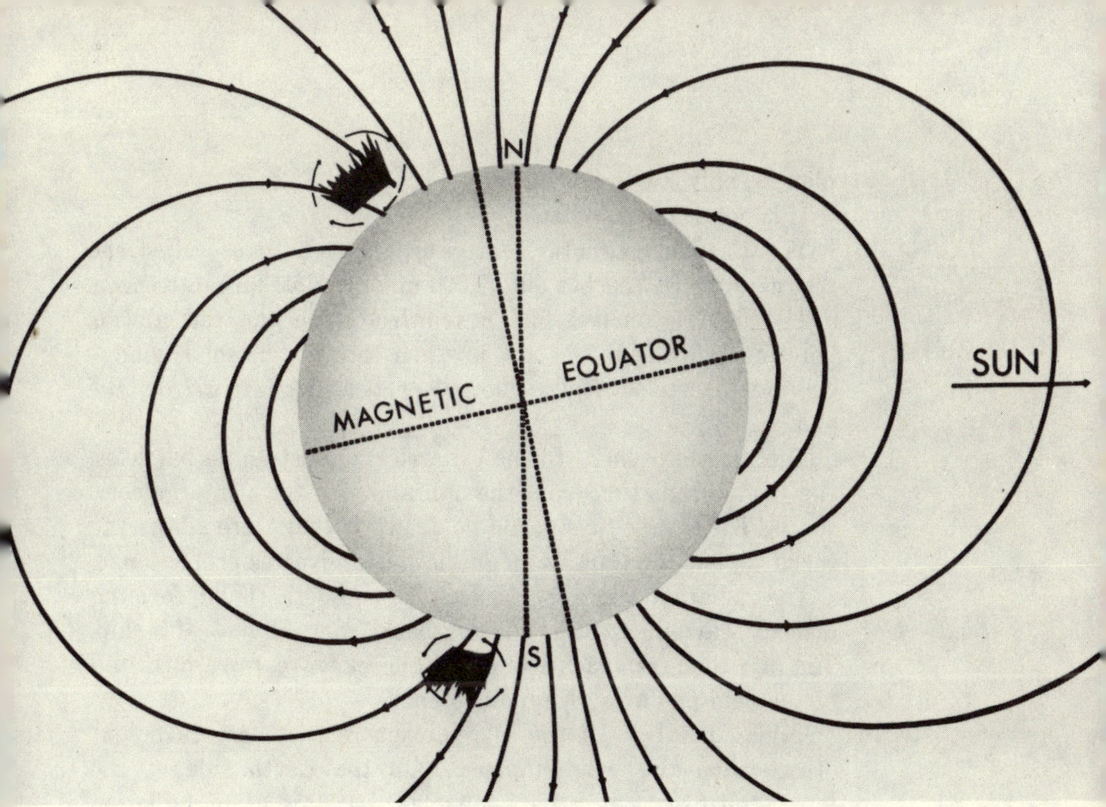

A

The magnetic field of the Earth is dipolar, which means it comes out of the top and travels down into the bottom—sometimes. Sometimes it does the opposite, but there is always a north pole and always a south pole, and the field helps protect us from deadly radiation

But then a physician, William Gilbert, using a magnetic rock called a lodestone, proved that the magnetic attraction came from the Earth itself.

Whatever it was, the power was reliable enough for men to stake their lives on it. If they only knew!

What would you say if we told you the magnetic north pole wanders all over the place? What if we said the north pole has sometimes been where the south pole is now and vice versa? That would make the compass seem somewhat less reliable than it appears. Well, it's true.

The Earth is not the only planet that is powerfully magnetic, but it is the only terrestrial planet we know of with that attribute. We know Jupiter has a strong magnetic field, and perhaps other Jovian planets do as well. But among the inner planets, only Earth has one.

It is not a minor matter.

The Earth's magnetic field, which is sometimes called the magnetosphere, reaches out 40,000 miles (64,000 km) into space.

The field is shaped like a teardrop with the tail always pointing away from the sun, brushed back by the solar wind—the breeze of electrically charged particles radiating from the sun.

The magnetosphere forms a barrier against those particles, the cosmic radiation from the sun and also from intergalactic space. The cosmic rays would be deadly if they were allowed to reach us, but the Earth's magnetic shield gives us protection.

The radiation comes in and hits the magnetic shield, forming a shock wave, like water pouring back from the bow of a ship. The rays then trail back around that shock wave, toward the tail of the field and away from our planet.

Behind the bow of the magnetosphere is a very turbulent area, called the magnetopause. On the Earth side of the magnetopause is an area known as the Van Allen radiation belts.

It had been assumed that if you sent a rocket with a Geiger counter into space from the Earth, the counter would show a rapidly increasing amount of radiation as the rocket soared into unprotected space.

But in 1958 the first American space satellite, *Explorer I*, tested that notion and found a surprise.

This tiny, 30-pound (13.6 kg) gadget was clocking radiation as it orbited the Earth. The orbit was a long, elliptical one, hitting an apogee, or high point, of 1,585 miles (2,550 km) and a perigee, or low point, of only 224 miles (360 km) from Earth.

Every time *Explorer* got to a point about 1,000 miles up (1,600 km), the counter reading went right off the scale. The machine went TILT!

Clearly there was something up there.

Dr. James Van Allen of the University of Iowa knew what it was. He said the magnetosphere was trapping radiation, preventing it from escaping. The trapped radiation formed belts around the Earth. Actually there are two sets of belts, divided by a "slit" about 8,000 miles up (12,800 km).

The belts have been named in Van Allen's honor. Once, when asked what practical value there was in his discovery, he replied amusingly that he has made a great living out of it. We now know enough about the magnetosphere to describe it fairly accurately. Envision a ball of material with a rod-shaped magnet inside it. Think of the classic schoolroom experiment in which iron filings are placed on a sheet of paper and a bar magnet is placed under the paper. The filings form a winglike pattern around the form of the magnet, looking as if they poured out of one end in both directions and merged to pour back in at the other end.

Earth's magnetic field is just like that. It is called a dipole,

The Earth's magnetosphere, buffeting up against the solar wind, the stream of tiny particles from the sun

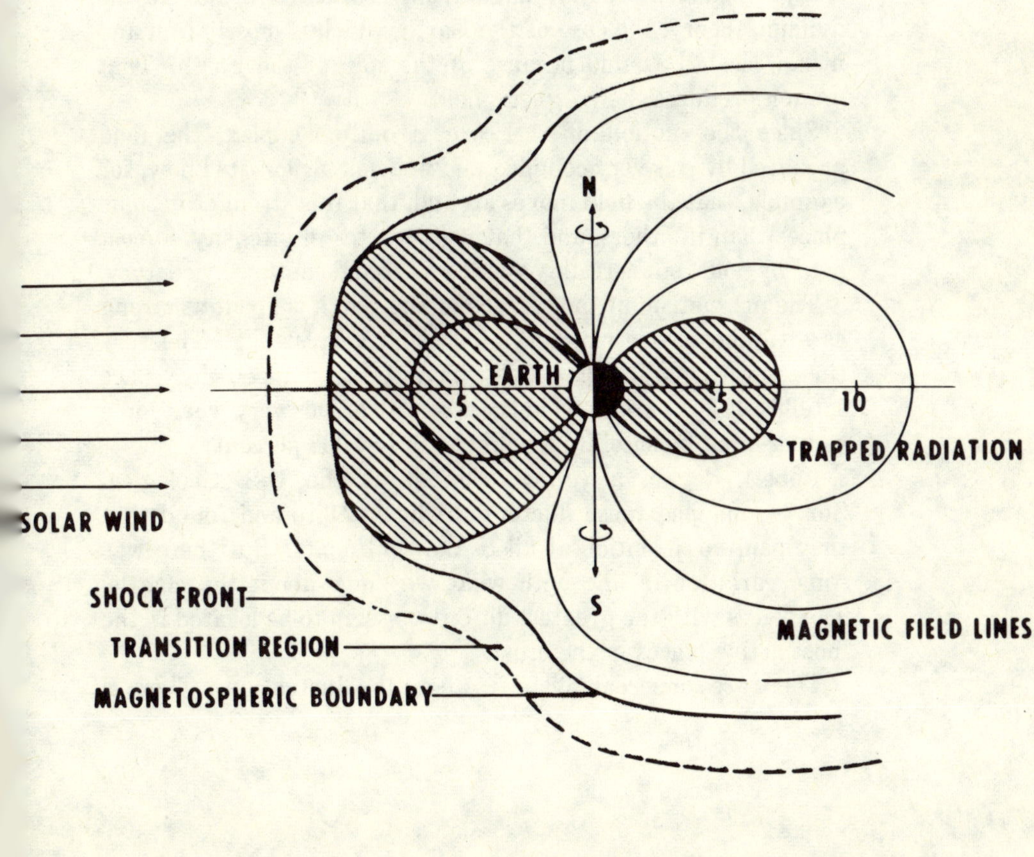

because it has two different poles, one positive, the other negative. To say it in a different way, it has a north pole and a south pole.

The only other magnetic field associated with a planet we know about is Jupiter's. Scientists believe that is also dipolar, but they think it is not straight. Imagine the arrangement of filings around a bar magnet with a bend in the middle. That is probably what the Jovian system is like.

The next obvious question is: Where does this magnetism come from? What causes it?

It's a good question for which there is a strong, but tentative answer. We are not sure, but we think we know.

In the 1930's two scientists, working independently, came up with what is called the dynamo theory. Walter Elsasser and Edward Bullard said that the liquid core of the Earth caused the magnetism.

Scientists know that if any substance with an electrical charge is moved around, magnetism results. According to the dynamo theory, the core of the Earth, which is mostly iron and nickel, sloshes around because of the spin of the Earth. That motion produces the magnetic field.

Since the motion of the core is quite complex, the field produced by it is very complex and uneven. Scientists know, for example, that the field moves around, that it is stronger in some places than in others, and that it fluctuates in intensity, almost hour by hour, but certainly year by year and century by century.

The magnetism at the surface of the Earth is about as strong as a toy horseshoe magnet, which isn't very strong. The very tiny variations in that field have been mapped by space satellites. The variations in the strength of this very weak force as measured by satellite amount to only 0.002 percent.

Robert Regan, a young scientist at the U.S. Geological Survey, mapped these fluctuations by satellite and found that they paid no attention at all to the continents. But there was some variation in line with what we know about the mantle. The places with the greatest difference seem to be located in the most active places in the crust.

There are some real mysteries about the fluctuations. There is

NASA

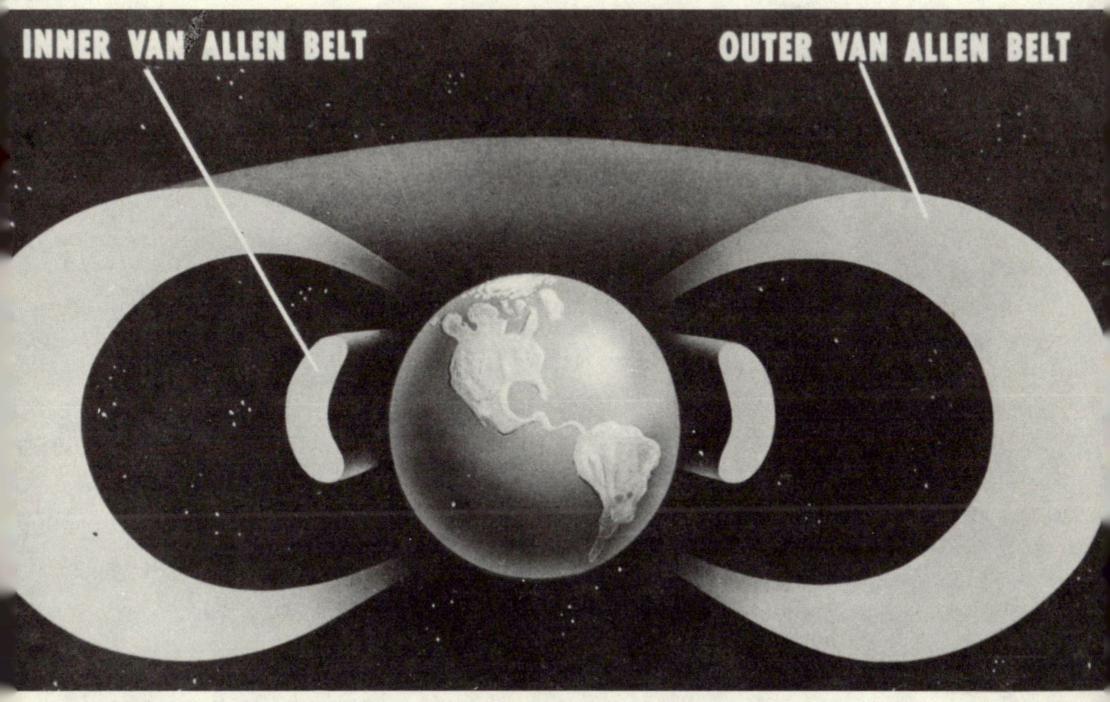

James Van Allen's famous belts of radiation, discovered by America's very first satellite. They come in two layers

a magnetic high that runs across the United States from the Pacific Coast to about Pennsylvania and then stops. There is no known reason for this high to exist.

There is a magnetic low that fills the Gulf of Mexico, and that is not understood. The Atlantic Ocean is full of highs and lows (unexplained), yet the Pacific is very dull (also unexplained).

Most of the measurements in the past were made by ships and planes, but now, with space satellites carrying very delicate equipment, we may be able to learn more about these fluctuations.

An interesting sidelight is found in the records of the first space satellites sent in orbit around the moon. Scientists noted they did not fly straight. They wiggled and wobbled, usually over the same place on the lunar surface. The reason, they have discovered, is that there are huge hunks of something highly magnetic beneath the lunar crust. These are now called mascons. The presence of these mascons has had to be taken into consideration for every Apollo moon landing.

One more thing. Can you guess how an explorer at one of the poles can guide himself with a magnetic compass and know when he is standing at the pole? He uses a vertical compass, one that points up and down instead of around.

The poles, we now know, move around. The magnetic north pole, now in Arctic Canada, roams all over the place.

How do we know that? Because the magnetic field leaves footprints behind.

When certain rocks like basalts are formed of molten material, tiny pieces of iron and other substances inside the rocks align themselves exactly according to the direction of the magnetic field.

The closer the rock is to the pole, the more vertical the alignment of iron in it. The closer it is to the equator, the more horizontal the pattern.

Such footprints are almost eternal. Once set in rock they stay there for millions of years, which is convenient if you are going to learn things from them. You almost have to tamper deliberately with the rocks to destroy the pattern.

Clays also preserve the magnetic pattern. Things like pottery keep the pattern of the field as it was at the time the pottery was made.

One scientist, S. Keith Runcorn, sampled rocks from all over the Grand Canyon of the Colorado and mapped the movement of the poles for hundreds of millions of years. He found that the poles wandered around through most of the Earth's history, the north pole sometimes even moving into the central Pacific.

About seven thousand years ago the wild roaming ended, and the pole now never moves more than eleven degrees from the geologic north pole.

But while reading the magnetic record scientists have found some very strange things. English scientists, for instance, found rocks in England that had a magnetic field that pointed in a different direction than other rocks of the same age found elsewhere in the world.

The only logical explanation was that somehow England had pivoted thirty-four degrees clockwise since those rocks were formed. That sounds pretty silly, the entire British Isles' turning, but as we shall see later it is not so silly as it seems. Those English rocks were just another clue in what was to be one of the most important scientific revolutions in the twentieth century.

But if you think that is weird, consider the rocks Bernard Brunhes found in the Massif Central, the mountain chain that runs through the middle of France. Brunhes' rocks were backward. In those rocks north was south and south was north, and nobody could figure them out. Perhaps, some thought, they were just odd rocks with something in them that had made them reverse their magnetic patterns, their polarity.

But soon other scientists discovered other rocks with reversed polarity, and it became clearer and clearer that those rocks demonstrated that at one time the Earth's magnetic field had flopped over. Other finds by other scientists showed such a reversal had happened more than once. In fact, the magnetic field reverses itself regularly.

We can say that about 700,000 years ago the field flipped to the reverse, and it has flipped back and forth dozens of times in the last 4 million years. It has, in fact, flipped 171 times that we know of in the last 76 million years.

If we consider the present setup normal (north is north and south is south), then the history of the Earth during the last 76 million years is that the magnetic field has been normal for an average of 420,000 years at a time, and has been reversed for an average of 480,000 years at a time. So, in fact, it is slightly more "normal" for north to be south and south north.

The periods of time are so similar that apparently there is no "normal." A visitor from outer space would be just as likely to find the fields reversed as to find them the way they are now.

The Earth's magnetism has been the way it is now for the last 700,000 years, which is unusually long—167 percent as long as average. We have found that only 15 percent of the polarity periods have been longer than that, although there have been a few that lasted as long as 3 million years. The shortest period was a mere 50,000 years.

Does this mean we are due for a reversal? Probably.

We think we know how this happens, even if we do not know why.

First the magnetic field begins to weaken over a thousand years or so. The field finally gets to a point where it is about one fifth its normal strength.

Then it seems to go mad, jumping around, back and forth, until finally the pole that was south settles more and more often to the north and stays there. The strength grows back to normal and what was once south is now north.

We are now in a period of decreasing magnetic strength, so that we may go through one of those magnetic flip-flops someday. It will not happen suddenly. The whole change can take four thousand years. But that sea captain's trusty compass is not so trusty as he might think.

What causes these events?

One theory is that the Earth gets smacked with a large meteorite from time to time and that bangs the field, causing a reversal. Also, objects from space called tektites, which are glasslike rocks, seem to strike land during magnetic events.

There is no proof yet that this is so.

One other odd thing. A large percentage of tiny sea creatures and some land creatures seem to have become suddenly extinct during magnetic events. This may or may not be cause and effect.

Some theories explain this, although none of them is totally accepted. One theory is that during an event the Van Allen belts weaken or disappear entirely, letting in radiation that would normally be blocked. The radiation would kill off some creatures.

Another theory is that the events produce violent geological disruptions that lead to sudden climatic changes, and that causes some creatures to disappear.

Still a third is that the core really begins to slosh, to become unstable, and when it "calms down" it finds its magnetic field reversed.

This is just educated guesswork. But if it is true that some entire species are destroyed by magnetic events, and we seem to be headed for one, are we in danger?

We just don't know. It does not appear to be anything worth worrying about at the moment, however.

Yet knowledge of this ability of the magnetic field to flip and leave its record indelibly ingrained in the rocks has proved to be very valuable.

It provided still another clue leading to the revolution in how we think about our world.

To understand that clue we must go to sea.

5
Down to the Sea in Ships and Such

> They that go down to the sea in ships, that do business in great waters; these see the works of the Lord, and his wonders in the deep.
>
> Psalm 107:23–24

"Request permission to dive."

"Roger, *Alvin*, you are clear to dive. Present water depth is 8,700 feet. Good luck."

The white, clumsy-looking vehicle begins to sink. Inside, the pilot and two scientists, crammed in a chamber 80 inches across (203 cm), look out of the window as the water begins to get darker and darker. The rolling of the sea they felt at the surface is now gone and the drop is smooth and quiet.

Soon it is completely dark except for the luminous sea creatures moving in order, like strings of lights on a night train.

The pressure grows on the 2-inch-thick (5 cm) titanium-alloy skin of the *Alvin*, the American research submersible. Soon the water will be pressing two tons for every inch of surface on the craft. If there is ever a rational reason for claustrophobia, a ride into the abyss in *Alvin* is it.

Even in the depths there are living things, scaled and shelled, scurrying out of the way of the floodlights that pierce the night of water.

At 6,000 feet (1,830 m) the pilot begins to move the ship

forward and *Alvin* now glides in a slow arc toward the Atlantic bottom.

At more than 8,000 feet (2,400 m) it buzzes over the floor of a fantastic geologic structure called the Mid-Atlantic Ridge.

It is the birthplace of the oceans and continents. Here the secret of the mountains and earthquakes was solved.

The *Alvin* and two French research vessels, the bathyscaphe *Archimède* and the diving saucer *Cyana*, designed by Jacques Cousteau, picked their way through the Ridge as part of an international exploration called FAMOUS, or French-American Mid-Ocean Undersea Study. They went there because it was becoming increasingly evident that only there were the answers to a large number of mysteries that had been bothering scientists for generations.

It was dangerous and exciting work.

As they glided along the floor the three men in *Alvin* saw lava that had bubbled out of the depths of the crust. Some of it formed piles, called "haystacks," 12 to 15 feet high (4–5 m) and 20 to 30 feet wide (6–9 m).

The lava flowed from the mantle of the Earth in tubes, beginning as molten rock 2,220° F (1,200° C) bursting into contact with seawater that was only 39° F (+4° C). The outside cooled and hardened, but the inside remained molten as more lava pushed up through it as through a straw in a milkshake.

The men aboard the *Alvin* saw some lava that looked as if it had been squeezed out of a toothpaste tube. Some looked like bulbs lying on the floor with piles of rock nearby, where the lava had pushed through the rocky crust. All the lava was going in one direction.

But in the middle of the Ridge, the submersible went over lava spreading both ways. They were at the very beginning of things, the Rift.

The dives regularly reach more than 9,000 feet (2,740 m). *Alvin* can work at depths of 12,000 feet (3,650 m) and a ship like *Alvin* has been tested to 22,500 feet (6,860 m).

Cyana operates down to 10,000 feet (3,050 m) and *Archimède* much deeper than that. It can withstand depths of up to 36,000 feet (almost 11,000 m) and can go anywhere in the ocean.

During one dive, *Alvin* became wedged in a crevasse and could not rise. Fortunately it was able to back up and work its way out or there would have been a real emergency.

Scientists consider work like this worth the effort and danger. Once men sailed atop the surface of the ocean and had absolutely no idea how deep it was or what lay beneath. That is not true any more.

In 1521 the explorer Magellan attached 12,000 feet (3,650 m) of line together and threw it overboard. It did not hit bottom, and Magellan figured he was over the deepest part of the sea. He was then in the South Pacific, near Tahiti, which actually is not the deepest part of the ocean.

Without electronic gear the bottom had to be measured in just such a physical way. You took a weighted rope of known length and threw it overboard, and when it stopped you measured it.

But usually it didn't stop. Sailors in the eighteenth century were forever failing to reach bottom. The American brig *Dolphin* once tried 39,000 feet (11,890 m) and missed the floor. A sailor named Parker on the *Congress* claims to have thrown 50,000 feet (15,240 m) overboard without hitting bottom.

What happened to the old line-and-sinker men was the current. It ran away with the end of the line.

One naval officer named Matthew Maury did not believe those figures, so he decided to try a scientific method, using sound waves. He clanged bells on windless days, hoping the sea floor would send back an echo. It didn't.

Maury went on to help develop more scientific methods and become one of the founders of the science of oceanography.

But by the turn of the century no one yet knew what was down there.

"The sea floor is a monotony," wrote scientist Alexander Agassiz. It is "only relieved by the dead carcasses of animals which find their way from the surface to the bottom and which supply the principal food for the scanty fauna found there."

A textbook during World War II reported: "From the oceanographic point of view the chief interest of the topography of the sea floor is that it forms the lower and lateral boundaries of the water."

Remember, scientists are frequently wrong.

Wars sometimes have good things associated with them, although the bad always outweighs the good. World War II produced electronic equipment that made it possible at last really to see what the topography of the sea floor was like.

One man who took quick advantage of this technology was Maurice Ewing of Columbia University. In 1947 he made the first real voyage using this electronic gear. His ship was called the *Atlantis*, and while Ewing was not looking for any lost continent, he found a lost world.

Maurice Ewing and one of his research vessels. He was a pioneer in studying the floor of the ocean from research ships

ont-Doherty Geological Observatory of Columbia University

Sailing from Woods Hole Oceanographic Institution at the base of Cape Cod, Ewing began to poke the bottom of the sea with an electronic sounder.

"I had the strange feeling of being an aviator flying high over an unknown planet tantalizingly hidden from view but outlined on his radar screen," he remarked.

Leaving Bermuda, *Atlantis* found the sea floor rising until it finally peaked at a sea mountain two miles high!

As it sailed on, the *Atlantis* made soundings by exploding TNT and recording the sounds. The scientists mixed the deadly stuff by hand on the tossing deck—dangerous work.

On July 28 *Atlantis* found itself over a flat plain, flatter than anyone had ever imagined a geologic structure could be. There appeared to be no obstruction higher than the height of a man. It was 17,400 feet (5,303 m) deep, and stayed at just about that depth for two and a half days of sailing.

This was the Abyssal Plains. They were first thought to be lava flows.

On July 30 the plains began to rise. Finally, at a place almost midway across the ocean, *Atlantis* discovered a place of wild mountains, canyons, rifts, and peaks—the Mid-Atlantic Ridge.

The Ridge is by far the largest mountain chain in the world, with peaks that make The Himalayas look puny and canyons that can contain several formations the size of the Grand Canyon of the Colorado.

Atlantis followed the Ridge south until she finally had to turn back. It was a great voyage of discovery.

We now know that the Ridge runs north and south through the entire length of the Atlantic and that there are ridges like it and connected to it in all the world's oceans. It is flanked on either side by an abyssal plain.

The Ridge runs 40,000 miles (64,000 km), making it the largest geologic structure on this planet.

The Ridge runs the length of the Atlantic, from Iceland south, curving around the bulge in West Africa and Brazil, staying right in the middle of the ocean, and going south again until it curves around the bottom of South Africa at the Cape of Good Hope. Then it goes into the Indian Ocean, where it splits.

GS

What Ewing would have seen if someone had pulled the plug and let all the ocean water out. Note the plains and, farther to the right, the Mid-Atlantic Ridge, and the rectangular patterns of rock on either side. There are formations like them in every ocean

One branch goes due north to the Gulf of Aden, then right up the center of the Red Sea until it finally becomes the Jordan Valley. The Dead Sea is a result of this feature.

The other branch turns southeast, going well below Australia and New Zealand, toward the Antarctic, and up the eastern Pacific off South America.

At the Gulf of California it hugs the North American coast, producing the deadly earthquake faults of California, where it seems to actually move onto land by San Francisco. Then it goes out to sea again, up the British Columbian coast, finally petering out near Alaska.

It soon became clear to scientists that this Ridge was alive. Some of the little knowledge gained about the ocean floor before electronic instrumentation came from cable companies that laid telegraph lines from North America to Europe. The cables broke frequently.

Some of the breaks were eventually traced to earthquakes. On November 18, 1929, there was an earthquake in the North Atlantic and twelve cables snapped. Six of them broke simultaneously, as was expected. All of those were within 59 miles (95 km) of the epicenter, the place on the surface of the seabed above the earthquake.

But the other cables broke in sequence, with those closest to the quake going first. More than thirteen hours later cables were still snapping. The last broke 300 miles (480 km) from the earthquake.

That is just one of the little puzzles scientists struggled with when they tried to study this dark, almost unknown realm. They were impelled to study the ocean and the ocean floor because it makes up three fourths of the surface of this world.

There aren't many ups and downs about our planet—what geographers call relief. The highest mountains are about 26,000 feet (8,000 m), and the deepest part of the ocean is 32,000 feet (10,000 m). The highest mountain of them all, Everest, is 29,028 feet (8,825 m), and the deepest "trench," the Mariana, is 36,198 feet (11,033 m) deep. If you took a pencil and drew a circle with a radius of one and one half inches, to represent the Earth, all the relief, from Everest to the Marianas Trench could be included in the width of that pencil mark.

Under the seas there are more to the continents than meets the eye. Offshore the sea floor is covered with shells and coral, teeming with life. It slants downward until, several hundred miles from shore, it meets what scientists call the continental slope.

In some places there is really a slope—a sort of gentle word—but in most of the sea the slopes would be better named palisades, for they are the highest, steepest cliffs in the world, towering a mile and a half tall (about 2.4 km).

The continental shelf and the slope are broken frequently by cuts and ravines and by gigantic canyons—many of them continuations of continental valleys cut by mighty rivers.

One of the largest canyons in the world, in fact, is the Hudson Canyon, carved through the shelf by the Hudson River.

Other canyons are not associated with rivers and some scientists think they were carved by glaciers during the ice ages when the shelf was exposed as dry land. The glaciers soaked up a great deal of the world's water and kept it frozen, so the sea level was much lower than it is now.

At the foot of the slope begins the continental rise—which isn't a rise at all if you go from landward—a much gentler area, usually flat but broken here and there by sea mountains. Some are high enough to stick above the ocean surface as islands. Most are covered by the rolling waters. Those that have tops worn round and smooth are called guyots.

Going from the land, the rise falls about 300 miles (550 km), dropping about 18 feet per mile (3 m per km) until it meets the Abyssal Plains, that incredibly flat structure that fascinated Ewing. Here the continental crust ends and the ocean crust begins.

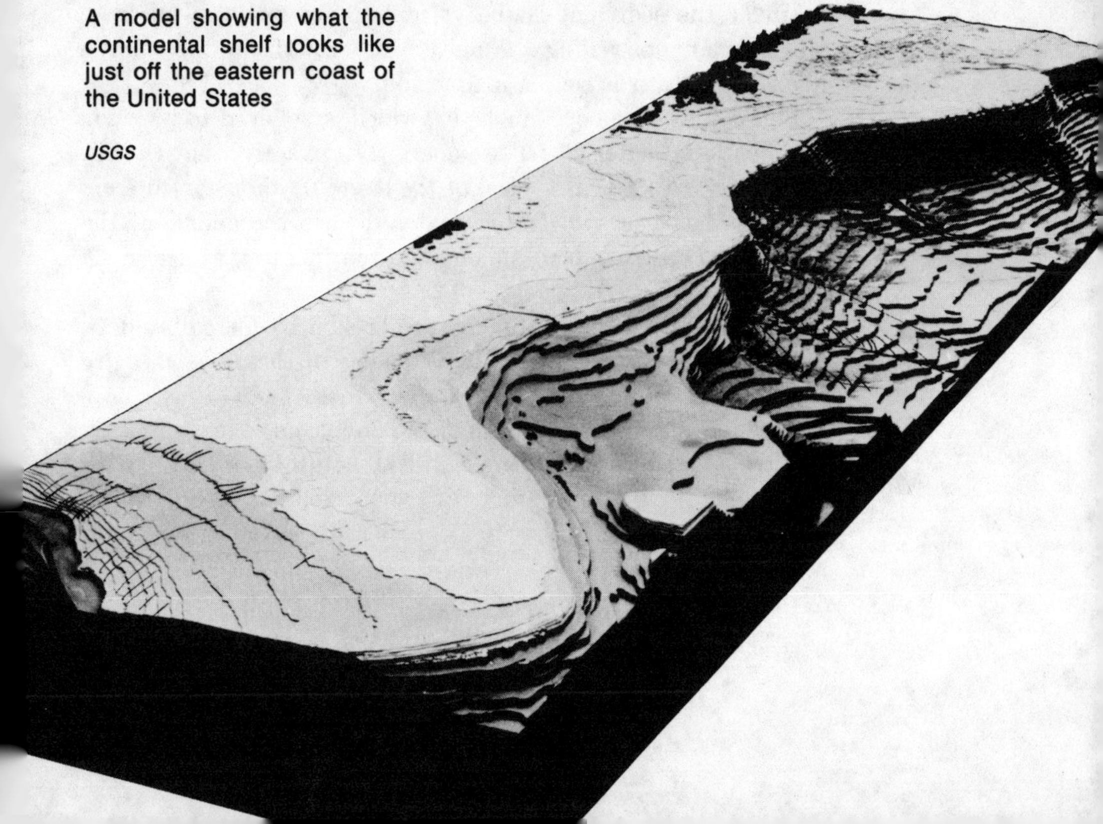

A model showing what the continental shelf looks like just off the eastern coast of the United States

USGS

There are some interesting things about this oceanic crust. For one thing, it gets thinner and thinner the farther out you go. At some places the Earth's mantle is only a few miles beneath the surface of the rocks.

The oceanic crust has three layers. The top layer is made up of sediment. One little peculiarity that turned into an important clue is that the sediment gets thinner as you go toward the center of the ocean and is thicker the farther away from the center you are.

Much of the sediment on the ocean floor comes from the continents, washed down by the rivers and by the tides. The rest of it is a product of the sea itself—dead, tiny microorganisms that sink to the bottom. But again, the sediment gets much thinner the closer you get to the ridges in the center of the ocean—as if it had been wiped clean.

Another odd thing is that the sediment is never very old—it is less than 135 million years old. Some of it ought to be much older, at least as old as the ocean. Scientists now think they know why this is so—and you will find out later—but until they pieced that little bit of evidence with the others, they were extremely puzzled.

Under the sediment is the volcanic layer made up of lava, which is an upswelling of molten material from the mantle itself. This layer is rich in magnesium.

Below that is the oceanic layer, which is believed to be made up of mantle material, but no one knows exactly what that is.

The picture so far is typical of the ocean if you traveled from the East Coast of the United States toward the middle of the Atlantic. There are places along the continents that are much different.

The edges of some continents are broken by incredibly deep trenches or sinks. On the bottom of some of these trenches are small plains like canyon floors.

These formations are typical of the Pacific and the Caribbean. One trench, the Puerto Rican, is 21,000 feet (6,400 m) deep, with one trench floor reaching down to about 28,000 feet (8,385 m). As we saw, the ones in the Pacific, such as those in the Marianas, are deeper still.

Remember those trenches. They are where the oceans die.

Even the Abyssal Plains are not devoid of life. Little scurrying things burrow through the sand, leaving ropelike marks behind them. There are fish at those depths and, believe it or not, great wealth.

Littering the floor are nodules or little balls of pure metals like manganese, copper, and iron. These minerals are part of the chemistry of the seawater. They have gathered in clusters around little objects like bones, teeth, or shells much the way an oyster grows a pearl.

A number of people, including the late, mysterious Howard Hughes, are trying to find ways of mining these nodules.

The plains are generally flat, but there are a few sea mountains. One chain, called the Kelvin seamounts, is found on the plain off the New England coast.

The Pacific has more mountains on its Abyssal Plains than does the Atlantic.

As you travel across the Abyssal Plains on your mythical journey you find the plains begin to rise, gently at first but then more steeply. Finally we are at the Mid-Atlantic Ridge.

It extends the length of the sea, split at about the middle by a rift, discovered by the famous oceanographic ship *Challenger* in 1950. The Rift is a deep valley cutting through the Ridge, the place explored at the beginning of this chapter by *Alvin.*

The discovery of this rift put scientists in shock. Many doubted it really existed.

There are similar structures on land. The best known is the Great Rift Valley of East Africa, a place where it appears the Earth has been torn apart. But scientists could find no justification for one under water.

Actually the rift was first predicted after scientists stumbled on another little mystery. Ewing and another scientist, Bruce Heezen, found that if you kept a record of marine earthquakes, you would see that most of them were located on a line that ran right down the middle of the Ridge. (For reasons we shall see later, the others were generally along the trenches, the other place in the ocean where the action is.)

The Ridge in the Atlantic is two miles high (3.2 km) and has mountains that soar 10,000 feet (3,050 km) above the plain. The Rift is 8 to 30 miles across (13–48 km) and is about 6,000 feet (1,830 m) below the mountains.

Not all of the Ridge is below the water. Iceland is a part of the Ridge that sticks up through the water, which makes it an excellent place to study this formation. The same is true of the Azores.

Another feature that puzzled scientists is the symmetry of the ocean floor. For instance, the formations on opposite sides of the Ridge are similar in many ways. If there is a long, flat plain on one side there is usually a long, flat plain on the other. Later on we will see just how spooky this symmetry is—the ultimate clue to Earth's mystery.

There are several other interesting features of the Ridge. There is, for instance, a large heat flux or flow in the Ridge. We have seen that it is also the site of many shallow earthquakes, which means it is geologically active.

Seismologists have found unusual wave velocities during earthquakes. That means that energy waves traveling through the crust do not behave the way scientists thought they ought to.

There is not only the lack of sediment to puzzle about, but there is the fact that there seems to be no erosion of the stark, thrusting peaks. Tides and currents erode just as rivers and wind erode. But at the Ridge the erosion is very slight, almost as if the Ridge was a new formation that had not yet been subjected to the millions and millions of years of eating away that created the guyots.

Finally, scientists discovered that the Ridge is the site of a very strong magnetic anomaly—a place where the Earth's magnetic field is stronger than usual.

We should also note that the Ridge is not a straight formation. It is broken and offset frequently by fractures and cracks that push whole sections to the side.

The faults all run perpendicular to the Ridge. They appear to push large square or rectangular sections of the sea floor away from the Ridge. Sometimes these fracture zones rub against zones moving in the opposite direction and the resulting friction causes earthquakes.

Some of the fractures are quite long, and the sections of the ocean floor show a kind of symmetry here too. The longer the fractures, the closer they are to the equator. The shorter the offset, the closer to the poles. The pattern is not exactly aligned along the pole of rotation and the real equator, but if you stick a rod through the Earth at a point somewhere near that pole you create another "pole"—a fictional pole called the pole of spreading, to which these features are perpendicular.

Amazingly you do not have to go diving to see what the ocean floor looks like at this place. There is an area, almost as impossible to get to, which appears to be a part of the ocean that forgot to acquire water.

It is called the Afar Triangle and it is one of the strangest, most important geologic sites in the world. Unfortunately, it is also the most miserable.

It appears to sit in the northeastern section of Ethiopia. Actually it is a French colony tucked in between Ethiopia and Somali. It is right across from the little bulge of Yemen that sticks out into the Red Sea at the Straits of Bab al-Mandab.

There may be as many as 150,000 people in Afars and Issas Territory (the formal name of the place), many of them fierce, wild nomads, armed to the teeth.

One scientist who goes there frequently, Haroun Tazieff, described it as "one of the world's most forbidding regions. In addition to the fact that its terrain is all but impassable, the area is extremely hot; we were to find that the temperature rises to as high as 134° F [57° C] in the shade in summer and 123° F [51° C] in winter.

"The region is inhabited only by nomadic tribes of fierce repute; the young warriors are said to mutilate male victims to offer trophies to their women, and they have been known to massacre armed parties for their weapons."

Into Afars go unarmed scientists with helicopters and Land Rovers (British versions of the Jeep). Why?

Afars is located about 575 feet (180 meters) below sea level. Three rifts converge here. One comes up from the Indian Ocean and is typical of oceanic ridges. One comes in from the Red Sea and is fairly straight, unlike oceanic ridges. The third rift is

NA

The Afar Triangle from above. This picture, made by *Gemini XI* astronauts in 1966, shows Ethiopia and Somali and the Gulf of Aden. The Afar Triangle is around the little indentation in Africa in about the middle of the picture. The stick on top is part of the spacecraft

part of the Great Rift Valley system that runs up East Africa.

All three rifts converge at Lake Abbe, a foul, wretched lake of bitter water.

Right through the center of the triangle is a trough cut by a series of parallel rifts. Molten matter oozes up from the bottom of these rifts forming dikes (walls) of rock.

The two sides of these rifts are being pulled apart by something and this permits the lava to make its way up into the cracks.

There are boiling springs and seven volcanoes. There is one mountain that may be a guyot, a peak such as we have described that was once underwater and had its top rounded off by tidal erosion.

The triangle appears to be rising, but scientists think it will sink again someday.

There is almost no water in the triangle except for that lake and occasional pools of salt water (salt is the principal natural resource of this awful place). No one knows why it is not underwater.

Tazieff believes that the Afars is typical of places where oceans are forming. It is also the place scientists must study to see how continents are created as geographic entities.

There is only one possible explanation for all the little mysteries scientists have found—the ocean floor is moving!

A scientist named Harry Hess decided that not only is the floor spreading, it is spreading from the Ridge, moving outward in both directions and finally sliding under the crust at the trenches.

There were problems. While that notion (or "model" as scientists call a hypothetical solution) seemed to fit the Pacific, which is ringed with trenches, it does not work quite so well for the Atlantic, which is not.

Hess, who worked at Princeton, published his concept in 1960. It has become one of the most important scientific papers of our time.

Let's go back and consider the ideas that led to this model.

The man who is considered the "father" of the science of geology is James Hutton, a physician fired by curiosity to understand the rock formations of his native Scotland. At the end of the eighteenth century he published a book now considered a scientific milestone: *Theory of the Earth.*

It was Hutton's contention that geology was a science of history and that the Earth as we see it now is the result of immense periods of time during which there was a slow, logical, explainable change. We are now fairly sure Hutton was right, although at the time he was considered somewhat radical.

The Bible was considered the core of historic knowledge in those days and it was full of sudden catastrophes, like Noah's flood—which, incidentally, has some historic support. The Flood, in fact, was an event that greatly shaped geologic thought. Geologic events were considered "antediluvian" if they occurred before the Flood, and "postdiluvian" if they occurred after it.

Most scientists believed in catastrophe as the shaping force of the world. If it wasn't the Flood, it was something else.

Sir George Darwin, as we said, believed that the moon ripped out of the Pacific—a sudden event that accounted both for the moon and the Pacific Ocean.

Alexander von Humboldt, a German scientist, believed the Atlantic was the result of a water storm or tidal wave that carved the ocean out of the crust.

Hutton's notion was the opposite of all this. No sudden catastrophe shaped the Earth, he suggested, only the slow evolution of rock.

Hutton did for geology what Charles Darwin did for biology. We call his approach "uniformitarianism." It is now the generally accepted geologic philosophy, but even to this day there are dissenters.

The most famous is Immanuel Velikovsky, whose book *Worlds in Collision* was a best seller in the 1950's and is still widely read on college campuses. Velikovsky explains many of the Biblical catastrophes, including the problems the Egyptians had at the time of the exodus, by the near-collision between the Earth and the planet Venus, which he thinks roared out of Jupiter one day.

Few scientists take Velikovsky seriously, but he retains a

Princeton University

Harry Hess, the Princeton geophysicist who explained the living oceans

popular, nonscientific following. So the catastrophe theory is hardly dead.

Hess was a uniformitarian, although he thought the Iron Catastrophe was pretty catastrophic. His idea of spreading sea floors, however, perfectly fits in with uniformitarian ideas, because it would permit scientists to answer many of the riddles of the sea (and of the Earth) without having to accept sudden events.

"Whole realms of previously unrelated facts fall into a regular pattern which is highly suggestive that close approach to satisfactory theory is being attained," he wrote in 1960.

Here is Hess's theory. Watch and see how it makes everything suddenly fall into place.

We have seen in Chapter 2 that Hess felt the ocean floor was not part of the crust but was really mantle. We have also seen how Meinesz explained the crust through convection currents.

Hess felt that the convection currents still rise through the mantle at the ridges.

The mantle is still pouring out through the rifts, in the center of the ridges, and is pushing the sea floor out in both directions. It is also producing volcanoes and the earthquakes Ewing and Heezen found.

The molten stuff hardens, as it pours out of the rift and accompanying volcanoes, and forms the sea floor. As more mantle pours out behind it, the hardened material is pushed away.

As the floor spreads, the continents are pushed ahead of it. Eventually the sea floor dives under some crustal material and is melted back into mantle—perhaps eventually to come out again as molten matter at another rift.

Hess identified the places where the floor descends back into the mantle as the deep trenches in the ocean, like those near Puerto Rico and the Marianas.

As the floor slides under crust, the guyots and mountains are scraped off. Many have been found near the trenches with a pronounced tilt caused by the sliding of the floor downward under the crust.

The reason there are no really old fossils and sediment on the floor is that the floor is constantly being renewed. The floor by the ridges has no sediment because it is too young. The floor by the trenches does have sediment, but before it can gather any really old stuff, it is shoveled under a plate of crust.

As the sea floor slides under the crust, the crust is misshapened and that causes mountains. This explains why most mountain chains are near the coast of a continent (or near what was once the coast) instead of running down the middle.

"The Atlantic, Indian and Arctic Oceans are surrounded by the trailing edges of continents moving away from them (as they are pushed by the spreading sea floor), whereas the Pacific Ocean is faced by the leading edges of continents moving toward the island arcs and representing downward flowing limbs of mantle convection cells," Hess wrote.

If there are earthquakes around the ridges and trenches, it is because something quite monumental is happening there, Hess said.

Hess was aided in his work by Robert S. Dietz, who pointed out as an example that large portions of the Pacific floor off the coast of California had apparently been moved vast distances relative to other parts of the nearby Pacific.

Hess and Dietz, who would provide ample evidence to back up their theories, had a little help. The ghost of a man you probably never heard of—one of the most important but least-known men in the history of twentieth-century science—was looking over their shoulders and grinning broadly.

Well, maybe he was crying.

It is time to introduce you to the main hero of all this. His name is Alfred Wegener.

6
Wegener

> What appears to be truth to the one may appear to be error to the other. And patience means self-suffering. So the doctrine came to mean vindication of truth, not by the infliction of suffering on the opponent, but on one's self.
>
> Gandhi, 1922

Some people become famous because they can explain the hidden. Few become famous because they can explain the obvious. Alfred Wegener explained the obvious, but he never became famous.

He is one of those tragic men who were either ignored or treated like fools during their lifetimes only to be proved perfectly sane and correct long after they are dead.

Alfred Wegener's crackpot scheme is now considered by modern scientists to be one of the most revolutionary advances in the history of science.

Wegener was born in Berlin in 1880, the son of an Evangelical preacher. His specialty was astronomy, and he received a doctorate in that subject in 1905.

He was something of a swashbuckler. He and his brother became fascinated with hot-air balloons and for a time in 1906 held the world's record—fifty-two hours. Along with this, Wegener had shifted his interest to meteorology.

The latter interest led to an invitation to join a Danish scientific expedition to Greenland. He wintered over there for two dark, frighteningly cold winters. This began a lifelong passion that eventually was to prove fatal.

He returned to Germany to teach at the University of Marburg.

We don't know how it happened, but it is not unreasonable to assume that one day Wegener looked at a globe and was puzzled. Everyone looking at a globe of the Earth is always struck by the same thing. But Wegener did something about it.

It seems odd that if you could slide South America right up against Africa you would find that the bulge in Brazil, near the city of Natal, fits snugly into the notch in the African continent near Cameroon.

In fact, it is obvious that all of what is called the bight (curve) in Africa at the Ivory Coast, Ghana, Dahomey, Nigeria, and down to Gabon, makes a remarkable fit with the entire northeast coast of South America at Brazil and Guiana.

It is almost too good a fit to be mere coincidence.

There are other places like that too. Look at the bottom of the Arabian Peninsula. It fits right into the East African coast at Somalia.

Scientists through the years would find even more remarkable fits when they dumped geologic data into computers and drew maps.

Wegener apparently thought this was all too remarkable. By 1910 or 1912 he had formulated the idea that the continents had once been part of one gigantic continent he called Pangaea, which means "all lands." The continent had broken up into parts which slid to their present places. That explained those odd geographic coincidences.

This idea did not exactly pop out of thin air.

In 1610 Francis Bacon saw the similarity and wondered aloud about it.

A few years later P. Placet wrote a little pamphlet called (from the French) *The Corruption of the Great and Little World, Where It Is Shown that Before the Deluge, America Was Not Separated from the Other Parts of the World.* That was 1668. Note the reference to the Flood.

In 1858, Antonio Snider looked at the globe and concluded

that all the continents had to be parts of one land mass. He called his tome nothing less than *The Creation and Its Mysteries Revealed*.

Around the turn of the century a geologist named Eduard Suess from Austria did more than look at a globe. He was struck by the geologic similarities found in the Southern Hemisphere.

He concluded that all of that hemisphere had to have been joined together into a supercontinent which he called Gondwanaland after a place in India that has produced interesting geologic samples.

There were two more who contributed to the notion of drifting continents, both Americans.

In 1908, F. B. Taylor had put together a theory similar to Wegener's. He claimed the continents had been pulled apart when the moon wandered by and was caught in Earth's gravitational pull (the "wandering stranger" model of Chapter 1).

Another scientist, H. H. Backer, in 1911 concluded something similar, but anticipating Immanuel Velikovsky, blamed the ripup on Venus.

Wegener's contemplation was interrupted by World War I. He served as a reserve officer in Queen Elizabeth's Grenadier Guards and was wounded twice.

While recuperating in the hospital for a wound in the arm, he started thinking about his theory again.

Even he was a little astounded by what he came up with. He wrote:

"The first concept of continental drift came to me as far back as 1910, when considering a map of the world, under the direct impression produced by the congruence of the coastlines on either side of the Atlantic. At first I did not pay attention to the idea because I regarded it as improbable."

By 1915 it was thought out enough for him to produce a book, *Die Enstehung der Kontinente und Ozeane*, or *The Origin of Continents and Oceans*.

It goes without saying that the general opinion of his contemporaries in science on the publication of *Origin* was that

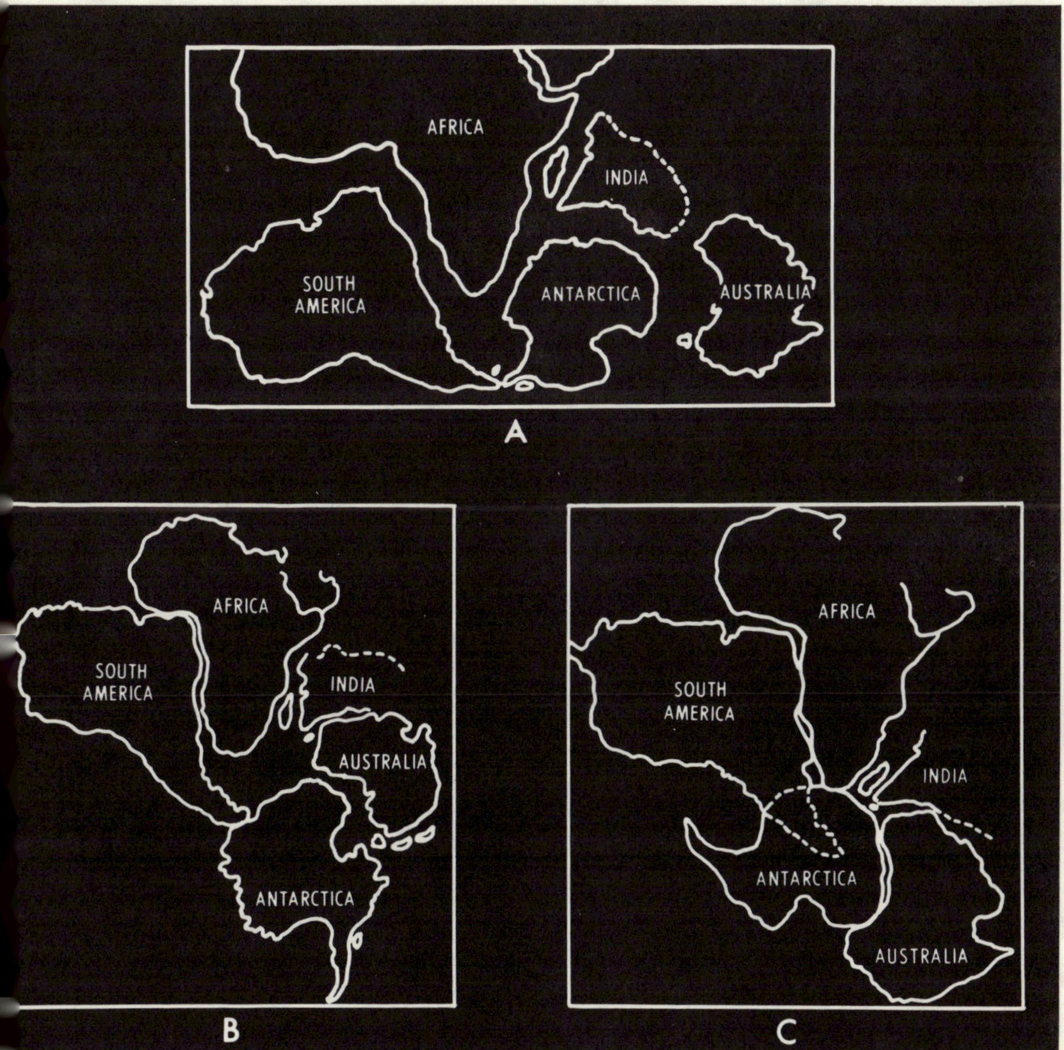

How the continents fit together. There are several versions

Wegener needed to have his head examined.

Wegener stuck to his guns, and in the following years put out several editions of *Origin*, each one adding more supporting data.

His theory was this:

All the continents were part of the supercontinent Pangaea until about 300 million years ago. Wegener showed with maps how the continents fit together.

The continents eventually began to break apart. Antarctica, Australia, India, and Africa had split off 150 million years ago. Africa and South America tore themselves apart later on and finally, only about a million years ago, Scandinavia, Greenland, and Canada came apart.

Iceland and the Azores, which we now know to be part of the Mid-Atlantic Ridge, were pieces of land left behind.

The great mountain chains like the Andes-Rockies that sit on the west coasts of the Americas, were formed when the plates of the continents moved. They are wrinkles in the Earth's crust.

Wegener was not entirely alone in all this. South African geologist Alex L. du Toit, using fossils and geology, produced considerable supportive proof of the theory in the '20s.

Both Wegener and du Toit found evidence that whatever happened geologically in South America happened at the same time in southern Africa, with unusual similarities. Glaciers did the same things at the same times. There are similar fossils, some of plants that were too similar to have popped up independently in two different places.

Some of the evidence was alive.

Wegener found great similarities between animal life in Australia and South America. For instance, those two continents are the only two with marsupials. Even islands near Australia do not have them, he noted.

He also found that hippopotamuses are found only in Africa and Madagascar (Malagasy Republic). That seems possible if you believe the two were connected, but is most improbable otherwise, because hippopotamuses can't swim.

The most telling evidence seemed to be plant fossils. Take, for example, the fern *Glossopteris*, a tropical, long-leaved plant.

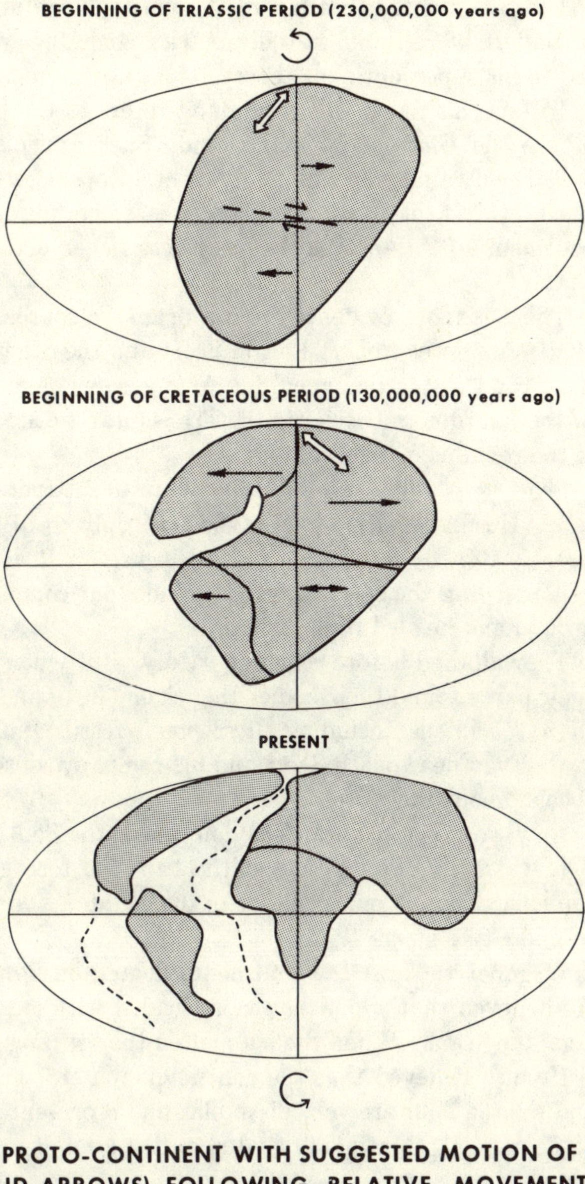

BREAKUP OF PROTO-CONTINENT WITH SUGGESTED MOTION OF CONTINENTAL PLATES (SOLID ARROWS) FOLLOWING RELATIVE MOVEMENT OF EARTH'S ROTATIONAL POLE (OPEN DOUBLE-ENDED ARROW).

USGS

The breakup of Pangaea, beginning 230 million years ago

When Suess noted that layers of sandstone containing *Glossopteris* flora were found in India, Africa, and Madagascar, he was led to his supercontinent theory.

In fact, his name Gondwanaland refers to the place in India where the *Glossopteris* was found. Suess, incidentally, never said Gondwanaland completely split into pieces. He thought most of it sank, and this part was something like the lost continent of Atlantis, although Suess never actually called it that.

There is more evidence from a dramatic source. In 1912 the British explorer Robert Falcon Scott and the other members of his second Antarctic expedition found seams of coal at the base of the Beardmore Glacier in the Transantarctic Mountains. Coal is the remains of tropical plants.

Scott was trying to be the first man to set foot on the South Pole. He managed to get to the pole only to discover he had been beaten by a month by the Norwegian Roald Amundsen. Still carrying the specimens, he and his four companions turned around and headed back to base.

They all died before reaching safety. But when the others in their party found their bodies they found 30 pounds (13.5 kg) of fossil specimens, including *Glossopteris* fossils embedded in the coal. Even near death Scott and his company sensed something special about their find.

Now what is a tropical fern doing near the South Pole? Well, if it wasn't the South Pole at the time the fern grew but was someplace more tropical, closer to the equator, you would expect to find *Glossopteris*.

Wegener thought the continents floated on lighter material. He believed that when they were loaded with ice, as in the Ice Age, they sank. When the ice melted they sprang up.

He also believed that the continents, out to the ocean shelf, and the sea floor are very dissimilar and represent two different kinds of rock. Continents floated on the kind of rock formation found in the sea.

Notice how similar that is to Hess's notion that the sea floor is really part of the mantle. Wegener seems to be right on target.

Wegener was also ahead of his time in another respect. He

felt that the ocean floor rock would have to be denser than the continental rock, but he couldn't prove it. There were no instruments available yet to test it, and it was at that time impossible to drill deep enough to bring back samples. Later technological advances proved his concept correct.

Du Toit helped with some geology. After a visit to South America he wrote: "Indeed, viewed even at short range, I had great difficulty in realizing that this was another continent and not some portion of one of the southern districts of the Cape [of Good Hope]." The rock formations were alike, even to the kind of formation that brings diamonds up from the bowels of the mantle.

Du Toit felt happier with two supercontinents. The southern he was content to call Gondwanaland from Suess. The other, made up of Greenland, Europe, North America, and Asia, he called Laurasia.

Wegener found some other geologic similarities. The formation that makes up the Appalachian Mountains is disturbingly similar to the Caledonian Mountains of Scotland and Norway. If Wegener was correct, they were formed when North America and Europe were joined.

Now how could continents move around on the Earth's surface? Scientists of the day were generally annoyed at the whole idea. It was widely assumed that the interior of the Earth was as solid as steel. Wegener suggested that like steel the interior became liquid at great heat.

He thought that at one time the world was covered with continental granite and that it was all underwater. He called the water the Panthalassan Ocean. Something started the continents moving, and they rumpled and crumpled until the ocean floor (mantle?) was exposed in places and the hollows were filled in with the remains of the ocean.

It is fair to say a battle broke out over Wegener's theory. Eventually it boiled down to Southern Hemisphere scientists (who saw enough evidence of continental drift to be sympathetic) against Northern Hemisphere scientists (who did not).

Wegener, however, kept plugging, adding new editions and accumulating more evidence and a few more allies.

He was "a man of medium height," according to his friend Johannes Georgi, "slim and wiry, with a face more often serious than smiling, whose most notable features were the forehead and the stern mouth under a powerful straight nose."

His theory made him so controversial he found he could not get topflight teaching positions. One of his problems was that he was the first to bring together knowledge from several sciences into one theory, just as modern geophysicists do now.

"One heard time and again that he had been turned down for a certain chair because he was interested also, and perhaps to a greater degree, in matters that lay outside its terms of reference," Georgi wrote.

It was almost as if he was not narrow-minded enough to fill a normal teaching assignment.

But the evidence kept mounting, and he never faltered. There is the fact that once there was an ice age in the Southern Hemisphere, with glaciers covering most of it. If the continents were where they are now, this would require more water than there is on the Earth to cover all that land. But if the continents were joined together, there is no problem explaining that ice age.

The scientific community largely heaped ridicule on the notion. The eminent British geologist Sir Harold Jeffreys wrote that the theory of continental drift "is a theory which explains nothing we wish to explain." Jeffreys kept tripping over Wegener's theory all his career.

One scientist reportedly foamed at the mouth when a colleague showed him a paper of Wegener's.

"Altogether," another scientist wrote, "the polemics [controversy] on continental drift are a sobering antidote to human self-confidence. Practically all the arguments against it, and many for it . . . now appear to have been fallacious."

Wegener was not around for the end of the argument. In 1930, at the age of fifty, he died during another one of his Greenland expeditions, probably of a heart attack.

His theory had gained some supporters but was generally considered more than a little weird. We now know he was not completely correct.

What he had done, however, was to announce a theory with enough evidence behind it to make it impossible to ignore.

When modern equipment and later scientists came to the fore, it was possible finally to settle the matter.

Wegener won—posthumously.

7 How to Move a Continent

A miracle is "anything with a probability of less than 10%."

Enrico Fermi

As is true of much of science, the evidence was all there. The theory merely required someone to come along and find it, and someone to put all the pieces together.

Wegener put the pieces together (in some cases incorrectly), but more and more pieces fell into line and the quantity and quality of the pattern became compelling, although the proof did not come until the mid 1960's.

You could look at the map and see how some parts of the continents fit with others. But how convincing is that?

Sir Edward Bullard, one of the originators of the dynamo theory that explains the Earth's magnetic field, knew that the real edges of the continents were not their shores—they were someplace under the sea.

Using a giant computer at Cambridge University Mathematical Laboratory, Bullard and his associates first asked the computer to find the best "rotation pole," a hypothetical place that would explain how the continents moved over the surface of the globe—if they did in fact move. The computer, using U.S. Hydrographic charts, found that "pole" and decided that the

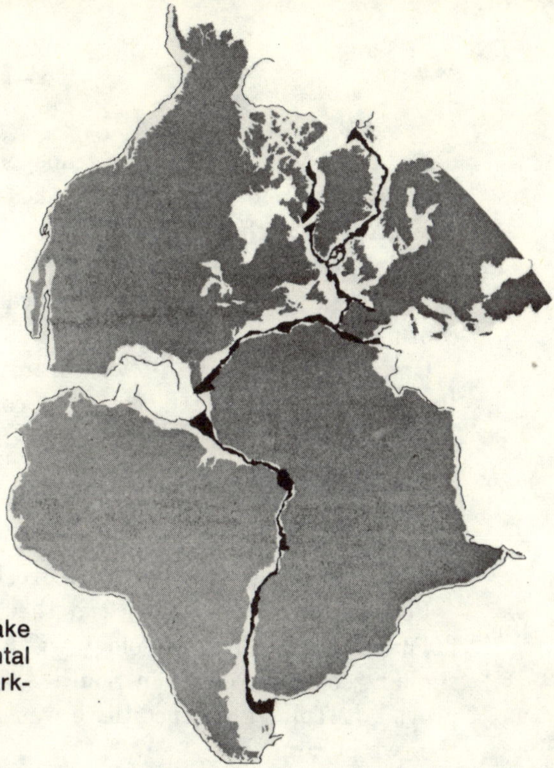

A computer fit. If you take into account the continental shelf areas, the fit is remarkable

USGS

best place to look for the real edge of the continents—if continental drift occurred, and if the continents really were joined once—was at the 500-fathom level.

A fathom is six feet, or 1.8 meters, so the rims of the continents were generally about 3,000 feet underwater (900 m).

Next the computer, using the 500-fathom charts, was told to fit the continents together the way scientists thought they might fit. The computer slid the continents back toward each other and found that the fit was almost perfect all over the world.

South America and Africa went hand in glove, as everyone expected them to. More surprisingly, North America, Greenland, and Europe snuggled into each other with ease. By turning Spain around a little a tight fit was found off the coast of Canadian Newfoundland.

There were some places that did not fit or that overlapped. The delta of the Niger River, near Port Harcourt, Nigeria, does not fit with the coast of Brazil, but scientists found that the delta, comprised of sediment transported by the river, is less than fifty million years old and the split between Brazil and Nigeria is older than that.

There are also some problems off the South African coast, but again these involve features that came about after the two continents drifted apart.

One place where the fit was dramatic was in the Red Sea. Backers of the drift theory were convinced the Red Sea is widening along the mid-ocean ridge that shoots up the middle.

The computer slid the coast of Africa and the Arabian Peninsula together and there was an almost perfect fit. The only place where a major problem appeared was, of all places, the Afar Triangle. The two sides overlapped.

Besides the visual evidence of the African–South American fit, modern scientific technology was able to match the ages of similar rock formations in South America and Africa, which proved du Toit's insistence there was something peculiar about the two places.

Even differences in the ages matched. There are two geologic formations involved on each continent, one two billion years old, the other 500 million years old.

When the computer moved the two continents together, the demarkation between the two formations matched exactly. The odds against that being a mere coincidence are fairly large.

Besides the coal and the *Glossopteris* fossils, there is another telltale formation, known to some scientists as the Dwyka Series, which appears only in the southern continents included in visualizations of Gondwanaland. These are thick, sedimentary rocks formed by glaciation which are so obvious scientists can plot the movement of the ancient glaciers from them.

In Britain there are rock layers that closely match rocks found in Greenland. Rocks in western Europe match those in the eastern United States and in Greenland.

One fantastic formation, called the Variscan orogeny (orogeny is the word for mountain-building), runs all the way from Alabama, through Newfoundland, Britain, Europe and down into northwest Africa.

The northern geologic evidence tends to support the notion

that the continents drifted apart, closed together again, and redrifted, instead of just splitting once.

There is evolutionary evidence besides the *Glossopteris* and the marsupials discussed earlier. There are some peculiarities in the family known as primates—apes and men.

Old World monkeys and New World monkeys are remarkably alike. It was often thought they represented a superb case of parallel evolution, the idea that two evolutionary lines, subjected to similar environments, would evolve in the same way even if they are in two separate parts of the world.

Some scientists found that difficult to swallow. One, Richard Hoffstetter, said a more logical explanation for the similarities would be a common ancestor. About 50 million years ago South America and Africa were much closer, and the primate ancestors must have managed to swim or drift between one continent and the other.

Problem—the age of the reptile lasted a whopping 200 million years. During that time evolution produced about twenty orders of reptiles (phylum, order, family, species). This was followed by the age of mammals, which lasted only 65 million years (and is still going on) but it has already produced thirty orders. Why? How is it that mammals are so much better at evolution than reptiles?

Answer: continental drift.

If, during the reptilian age, the reptiles were confined to only two continents (Laurasia and Gondwanaland) they would have had less opportunity to evolve than the mammals who rose after the continents had split, according to Bjorn Kurten.

There is a kind of evolution called adaptive radiation. This is when a species enters an environment that has a number of adaptive niches—differing environments that permit it to evolve in differing ways.

One example is the finches Darwin found on the Galápagos Islands. The first finches came over from the South American mainland, probably by getting lost. They spread around the islands, and because finches are not very social, they stayed on their own islands and evolved separately.

Eventually, where they had all been one kind of finch, there were now thirteen.

The opposite of this is called convergence. In this type of evolution a number of widely differentiated species evolve toward one another, gradually becoming quite alike. In mammals this is found among anteaters.

Reptiles, roaming two large continents, would have fewer niches in which to adapt. So they evolved into fewer orders. On the other hand, mammals were segregated on a half dozen continents, each in a different place with differing environments. Adaptive radiation would have provided for more orders in a shorter period of time, Kurten maintained.

His conclusion was that "we ourselves, as a subgroup within the order Primates, probably owe our origin to a radiation within one of Gondwanaland's island continents," probably Africa. From Africa the early hominids (human ancestors) spread to Asia and Europe.

"The cycle of continental fragmentation and amalgamation thus seems to have played an important part in the origin of man as well as of the other land mammals."

But doubters found the fossil evidence inconclusive. Even Scott's hard-won fossils did not impress them completely. It was possible, they maintained, that seeds or pods had drifted in the wind to Antarctica from elsewhere. Certainly over a few million years there was ample opportunity for this to happen.

They maintained they would not be convinced until somebody found a land animal's fossil in the Antarctic.

During the International Geophysical Year 1957–1958, two scientists working in the Transantarctic Mountains discovered what was apparently part of a reptile's jawbone. The reptile involved is called a labyrinthodont, a squat, snub-nosed creature not far removed from the first amphibian.

Almost, but not quite good enough, said the critics.

At the end of 1967, two scientists, Edwin Colbert of the American Museum of Natural History in New York, and David H. Elliot of Ohio State University, clung to the sides of a bluff near the Beardmore Glacier in the Antarctic in temperatures that would have driven an Eskimo indoors. Their faces were so frostbitten that they all broke out into open sores when they finally returned to base.

But they pulled out of the bluff unmistakable reptilian bones 200 million years old. The find was similar to fossils dug up in South Africa.

A few years later came the clincher, the almost complete skeleton of a Thrinaxodon, a reptile that looked something like a weasel, also 200 million years old. This discovery by James Collinson of Ohio State—an extraordinary piece of luck combined with heroic perseverance—was the first of several such finds in the Transantarctic Mountains.

Continental drift seemed inevitable.

But the most telling evidence of drift would come, of course, from the source, the ocean.

We have already seen that there are some strange things going on down there. Unraveling them was impossible until modern technology developed.

Perhaps the greatest example of this technology is the research ship *Glomar Challenger*. *Glomar* refers to the Global Marine Company, which operated it. Challenger goes back to the H.M.S. *Challenger*, the nineteenth-century British oceanographic vessel.

The new *Glomar Challenger* is a weird-looking vessel, something like a freighter with an oversize oil derrick on top. It is capable of drilling to depths never explored before, using a bit 9.25 inches in diameter (23.5 cm). The drill system returns core samples of the ocean floor to the scientists on board.

It is a rather drastic way to search for fossils. It is also a fine method of dating the sea floor.

The trick to this kind of drilling is to keep the ship exactly over the drill hole, come waves, tides, or wind. *Glomar Challenger* has four propellers for maneuvering. Each is connected to a computer.

When the ship gets over the drill site a special acoustic beacon is thrown overboard. When it hits the bottom it sends off acoustic signals to hydrophones (underwater microphones). The length of time it takes the signal to get from the beacon to the hydrophone tells the computer exactly where the ship is in relation to the beacon (and the drill hole), and it signals the propellers to compensate for any change in the ship's position.

In 1968 the *Glomar Challenger* was used to test the theory that the ocean floor was in motion.

It sought to drill through the sediment across the Mid-Atlantic Ridge. The ship failed to get deep enough on a number of holes. But at one series of holes, at about latitude 30° south (which runs through South Africa and southernmost Brazil), the *Glomar Challenger* brought up fossil records that appeared to confirm the idea that the sea floor was moving—it appeared to be moving away from the Ridge on both sides, probably at a rate of about 1.6 inches (4 cm) a year.

Many years earlier, Dutch geophysicist Felix A. Vening Meinesz took a gravitometer (an instrument that measures gravity) down in a submarine over the trenches. He recorded the biggest drop in gravitation ever recorded on Earth. Something was pulling the crust downward at the trenches and it wasn't gravity.

Arthur Holmes and D. T. Griggs went over the data and suggested that the answer was in convection currents. At the trenches the currents were heading back down into the mantle and they appeared to be dragging the crust down with them. The currents come up at the ridges and go down at the trenches.

And all of this left footprints behind—clues that lead to the inescapable conclusion that continental drift is real. The footprints, of course, are magnetic.

We have seen how rocks, when they solidify, retain the magnetic images they have acquired. We have also seen how the Earth's magnetic field has reversed itself scores of times. Those are the two main features of the solution.

Every field reversal is called an anomaly or event. Scientists have begun to find some peculiar things about the oceanic anomalies.

Research ships have crisscrossed the ocean, measuring these anomalies. Ronald G. Mason and Arthur Raff of the Scripps Institution of Oceanography, using shipboard magnetometers, discovered that the ocean floor was magnetized in stripes.

In 1963, F. J. Vine and D. H. Matthews of the University of Cambridge suggested that the magnetic anomalies in the

Atlantic ridge area were the result of magnetic field reversals. Vine and J. Tuzo Wilson of the University of Toronto, for instance, had found surprising symmetry in the magnetic pattern on both sides of a ridge near Vancouver.

The men suggested in the magazine *Nature* that "if spreading of the ocean floor occurs, blocks of alternately normal and reversely magnetized material would drift away from the center of the ridge and parallel to the crest of it."

This is apparently what happens. First, molten rock comes up through the ridge and solidifies. This rock we will call Rock A. As it solidifies it acquires the magnetic trace of the field in existence at the time. Let's say the field when Rock A solidifies is normal.

The field reverses. Now rock coming up through the rift, Rock B, solidifies with reversed magnetic traces. As it pours up, it pushes Rock A away from the rift in both directions. We now have Rock B at the ridge with reversed magnetism, and on either side Rock A with normal traces.

The field reverses again and the new rock, Rock C, takes on normal magnetic traces, pushes Rock B out in both directions, and pushes Rock A even farther away from the ridge.

Now, looking down from above, which is what happens when you sail over it in a boat, you have Rock C at the ridge with normal magnetic traces, bordered on each side by a strip of Rock B with abnormal magnetic traces, and on each side of B, strips of A, with normal traces. That is how you get stripes.

Vine and Matthews had trouble selling this notion at the time. But one scientist engaged in similar research, James R. Heirtzler, then of Lamont Laboratory of Columbia University (Ewing's outfit) reported: "By 1965 it was clear to us and others that magnetism could be the key to reconstructing the history of the ocean floor and the movements of the continents."

Lamont scientists also found that every time there was a fracture along the ridge the stripes moved along with the dislocated rocks and kept their pattern completely.

Since we know the dates when the fields reversed, we can measure how fast the sea floor is moving by measuring how big a particular stripe is.

The clincher was a piece of computer paper called *Eltanin-19*. *Eltanin* was a National Science Foundation research vessel and 19 was the voyage number. The ship took pictures of the magnetic profile as it sailed and recorded them as a series of lines on computer paper. A whole profile looked like an erratic graph with hills and valleys squiggling all across the paper as the ship went from one place on the ocean to another.

Eltanin-19 looked just like the rest of them at a quick glance, but in 1966 a graduate student, Walter Pitman at Lamont, looked again and saw something spectacular.

The "ups" and "downs" on the chart were completely symmetrical. If there was an "up" of a certain size on one side of the ridge, there was an "up" of the same size in the same place on the other side of the ridge. There were no interruptions in what turned out to be a perfect specimen.

Pitman was able to ask the computer to produce a graph showing what the magnetic profile would be expected to look like if ocean spreading was a fact. The computer came up with a graph that looked exactly like the one *Eltanin* found on voyage 19.

Scientists were beginning to match up stripes with known reversals. One by one they seemed to match what research ships were finding.

The mirror images were so identical on some profiles that it was possible to take two slides, one of each side of the ridge, flop them over and hold them one on top of the other. They would match perfectly.

Everyone stopped laughing at Vine and Matthews.

By tracing all the ridges, and the apparent spreading, it was concluded that the continents appeared to be sitting on huge plates of rock—tectonic plates, they are called—being pushed around by the sea floor as it spreads from the ridges. A trench is created when one plate collides with another. One of them has to give way and tucks under the other—like a global game of chicken.

When one plate meets another, the one on top frequently crumples, forming mountains like the Andes. The plate that takes the dive carries mountains and guyots with it, until the top plate scrapes them off. The sediment is also removed, which explains why there is no such thing as sediment as old as the ocean.

Around the trenches there are earthquakes and volcanic action.

One puzzle was how underwater volcanoes can move along with the spreading sea floor. They pop up at the ridge and grow higher as they are carried away by the moving plate.

The lava comes from various magma chambers—huge rooms of molten matter in the mantle. As the volcano slides along it just keeps finding new sources of lava.

How fast does the floor spread? It varies between half an inch and a little more than two inches (about 1.3 to 5.1 cm) a year. There is no evidence the spreading ever stopped to rest for a while.

What about Wegener's theory of one supercontinent, Pangaea? If it existed, it was not long-lived as these things are measured. It lasted perhaps only about 50 million years.

Two scientists, Robert Dietz and John Holden, have reconstructed Pangaea. They used the continental slope as their margins, the 1,000-fathom (6,000 ft. or 1,830 m) mark.

They admit to having some problems fitting the continents together. For instance, Florida and the Bahamas do not fit when you slide North America over northwest Africa—one of the reasons some scientists are skeptical about Pangaea. Dietz and Holden explain those problems by saying that this land was created after Pangaea split up.

"According to our reconstruction," they wrote in *Scientific American* in 1970, "Pangaea was a land mass of irregular outline surrounded by the universal ocean of Panthalassa, the ancestral Pacific. The fit between North America and Africa provides the principal connection between the future block of northern continents and the future group of southern ones."

Between what is now eastern Africa and Arabia and Asia is an ocean called Tethys, now almost completely extinct. Scientists are pretty sure Tethys existed for hundreds of millions of years as late as 200 million years ago.

The Mediterranean is all that is left of Tethys. There is

another gap between Asia and Alaska which became the Arctic Ocean.

They figured that when measured down to the 1,000-fathom line "the total area of Pangaea was 200,000 square kilometers [80,000 sq. mi.], or 40 percent of the Earth's surface—equal to the area of the present continents measured to the same line."

What caused Pangaea to start breaking up? Molten material poured through rifts, Dietz and Holden say. There were two rifts in Pangaea—one opened the Atlantic, the other the Indian Ocean.

No sooner did Pangaea break up than Gondwanaland and Laurasia started to split, they said.

North America slid toward the northwest, opening up the new Atlantic Ocean until it was about 620 miles (1,000 km) wide. Dietz and Holden think at that stage the Atlantic and Pacific were probably joined somewhere. The eastern coastline of the United States then ran east-west, with coral reefs growing up and down the shore.

A rift opened from the young Atlantic, splitting Greenland away from Labrador (Labrador Sea).

The Tethys Sea began to close. Farther to the west, Spain rotated, opening the Bay of Biscay.

The first rift between South America and Africa occurred, with a line that ran up as far as where Nigeria is now, forming a long, narrow sea.

The South Atlantic later began to widen, causing the final separation between the two continents. The ocean grew to about 1,900 miles (3,000 km) wide.

Africa went north and started rotating clockwise, which almost closed the Tethys. Antarctica and Australia, still joined, rotated westward.

North and South America, riding on separate plates, moved westward until the plates finally hit a trench in the Pacific. The folding caused by this meeting led to the first stages of the California coast mountain ranges in the north, and the Andes in the south.

About 65 million years ago the continents were more or less in the position they are in now. The Atlantic kept widening and

Greenland split from Europe.

There were three final, cataclysmic events. Volcanic action built up the Isthmus of Panama, joining the two Americas. Australia and Antarctica finally broke up.

And, in a major geographic collision, India, riding on its own tectonic plate, collided with Asia and its plate. The result was that Asia succumbed to the force, part of it rumpled violently, and the result is called The Himalayas.

This version is now widely accepted, but there are others. Some scientists think there were two Pangaeas—that the continent broke apart, clanged back together, and then went through a split similar to the one just described.

The first supercontinent is called Pangaea 1 in this model and it existed until about 450 million years ago. It was located mainly in the Southern Hemisphere. The Gondwanaland continents were together but North America and Europe were not in their Laurasian configuration.

It was not a nice place. There were no plants, just a baking sun in the summer and frozen winds in the winter. In the places where the rifting was to take place there were tall cliffs and rift valleys. There had just been an ice age, so the rocks had been scraped by the moving mass of ice. There was some sedimentary rock, but generally the rocky floor was crystalline in nature.

Growth along ocean ridges went so fast that it might have shrunk the oceans' capacity to hold water and made the water splash onto the continents.

About 450 million years ago Pangaea 1 began to split up. Volcanic action heralded the divorce and all hell probably broke loose over what was once a very quiet place. As the land mass split, new oceans spilled in between the gaps. Sediment where the earlier seas had been was left behind along the edges of the continents and in a few interior places. Scientists believe the equator during this time could have been drawn from Central Europe through Greenland and on a line to California.

The volcanic action remained fierce, with lava flowing out onto the surface to form the great lava shields that fill much of the continents. The worst action was along the coasts—California to Alaska, Canada to Europe.

The southern continents were probably quieter, although scientists have little information to go on which would enable them to describe what happened there.

About 400 million years ago, we know, there were many shallow seas over the continents, all of which left their sediment behind them.

Between 350 and 400 million years ago, for reasons unknown, the continents started drifting back together again. By this time vegetation had covered much of the land and the continents were somewhat less miserable.

Some land rose out of the seas, including a minicontinent in what would be the North Atlantic. Mountains crumpled up, including the Appalachians, which are some of the oldest mountains in the world—once impressive towering peaks that would make the present Rockies seem puny in comparison.

It rained a lot, and the water began to run in rivers and streams, moving rock around with it in the form of sediment.

Gigantic forests grew that would eventually form the coal beds that fired the Industrial Revolution. In North America, according to David Dinely, the forests probably covered about 160,000 square miles (414,000 km^2) and in Europe about 62,000 square miles (161,000 km^2). Much of the plant life was the *Glossopteris* that is now so famous.

As the continents slid back toward each other, the masses that were to form Laurasia crushed together violently. More land was heaved around, and more volcanoes, the self-appointed trumpeters of geologic happenings, blasted through the land.

Among other things, this dried out most of the seas and the Laurasian continents, except for some parts of the Arctic.

This also killed off the coal forests and brought deserts to where lush greenery once thrived.

Again, most of the action was in the north. The Gondwanaland continents were much quieter. They were generally a lush place. Dinely describes them as being full of streams and vegetation, coral seas teeming with life.

About 350 million years ago an ice age hit Gondwanaland. But those places not covered by masses of ice were covered with cool, wet forests of *Glossopteris* (a different kind than the other *Glossopteris*). It probably looked something like the Pacific Northwest and the Olympic forests today.

One thing about glaciers—they move around a lot of land, scraping the top of the rocks and pushing the scrapings before them as they slide across the countryside. They also scour ravines and canyons, tearing the rocks with their motion and weight.

Eventually the land met again and this area is called Pangaea 2. From here the Dietz and Holden model would pick up.

Well, we are now convinced the continents move around. How do they do it?

It is obvious that continents are very large and weigh a great deal.

First of all, scientists had to rethink their conception of the mantle. It certainly cannot be rigid.

Lord Kelvin, a famous British scientist of the nineteenth century, had figured out that because the Earth bent so little under the tidal influence of the sun and moon, it must be more solid than steel.

Early geologists, however, were confronted with lava, molten rock that flowed like glop out of volcanoes. It would seem logical to assume that this stuff came from below the surface and that it was not buried very deep, probably in the upper mantle.

It also seemed clear the crust was floating on something.

Any floating body seeks an equilibrium. Heavier stuff sinks. Lighter stuff floats higher. This seemed to be the case with continents, based on studies of gravity in the nineteenth century.

If something floats, then the material it is floating on must be denser. So it was clearly established that something below the crust (the mantle) consisted of denser material. Also, for the floating material to adjust to changing weight, the underlying material must be somewhat plastic—in other words, have the capacity to flow and change shape.

When a continent goes through an ice age, the weight of the ice bears down upon the crust and the continent sinks a bit. If the Antarctic ice shelf melted, the continent below that ice would pop up about a mile.

Other continents have gone through less drastic motions. The land covered by the last ice age is still rising about a meter (three feet) a century.

If the mantle flows, it flows very, very slowly.

The key to proving Kelvin wrong came from listening to earthquakes, somewhat the way a doctor listens to a patient's heart with a stethoscope. Earthquake waves act differently when traveling through different kinds of rock. By putting together seismic records of earthquakes, scientists were able to put together a picture of the interior of the Earth that made it seem likely the mantle was indeed plastic.

According to geophysicist Don Anderson "the tone of a bell—that is the frequencies at which it vibrates—depends on its elastic properties; a steel and a bronze bell emit different sounds.

"From records of the free vibrations following a big earthquake it is possible, with enormous mathematical labor, to deduce the elastic structure of the Earth. The labor has been performed."

The conclusion is a plastic mantle.

That accounts for the seemingly improbable concept of continents plowing through the mantle like steamboats through oatmeal. But what makes them move?

It is now believed that convection currents are the engines of continental drift.

When the research ships were scouring the seas with magnetometers and gravitometers, they were also bringing thermometers.

The first clue came in the Red Sea. Scientists are convinced that the Red Sea is spreading, pushing Africa and Arabia apart. *Apollo 17* astronaut Harrison Schmitt, a geologist and the first American scientist in space, looked down at the Red Sea and told ground control: "I didn't grow up with the idea of drifting continents and sea floor spreadings, but I tell you, when you look at the way the pieces of the northeastern portion of the African continent seem to fit together, separated by a narrow gulf (the Red Sea), you could almost make a believer out of anybody."

In the late nineteenth century, scientists probing the depths of the Red Sea began to find that there were some odd temperatures being recorded at the sea floor.

In 1963, the British research vessel *Discovery II* (named after Robert Scott's Antarctic vessel) found four large pools of hot brine on the floor of the Red Sea. The largest was eight miles long (13 km) and three miles wide (5.5 km) at a depth of 6,800 feet (about 2,100 m). The temperature of the water was a phenomenal 133° F (56.1° C). It was very salty, 317 parts per thousand or about ten times normal.

Peculiarly, there was no dissolved oxygen in the water, as there usually is in seawater, and the sediments under the pool were full of minerals, like zinc, lead, and silver. The crust under the sediment was unusually hot.

The pools were so dense that instead of rising, as hot water will in cooler water, they just sat there on the bottom.

This is apparently where the spreading is taking place, and the heat from the underlying lava is creating the brine pools.

Other pools were found by the Woods Hole ships *Atlantis II* and *Chain*, and one of those was even stranger. It was getting hotter.

In twenty months the temperature went up one degree Fahrenheit (about .6° C). This could, of course, be just an error in measurement, but later tests showed the temperatures still climbing. A lot of hot water was pouring into this pool, more than could be found at a Yellowstone geyser.

Incidentally, the sediment found by *Atlantis II* was also rich in minerals, and someone once figured out that the ore deposits under that brine would be worth about 2.4 billion dollars under 1968 smelter prices. First, however, you would have to figure out how to get it.

The only logical explanation for these hot spots is convection currents in the mantle—heat rising from the interior of the Earth.

But does that drive the plates and push the ocean floor? Scientists are still unsure. The most attractive theory at present is that convection currents *are* the source of the movement.

W. Jason Morgan of Princeton believes there are about

twenty places around the world where heat carried by hot rocks is pushing up from the mantle. The heat gets to the crust and spreads out under it. This pushes the continental crust, sort of like a conveyor belt.

Morgan called these flows "plumes."

There are six large plates and several small ones, according to the theory of continental drift. They are called tectonic plates because "tectonics" is a reference to the Earth's crust, and the whole concept is known as plate tectonics. Make a note of that.

There is the North American plate, which includes all of North America and Greenland except for a strip of California. We will see later what a problem that causes.

The South American plate contains all of that continent and runs, like the North American, to the Mid-Atlantic Ridge. It hits the Nazca plate (one of the small ones) on the west.

There is an African plate running from the Mid-Atlantic Ridge east to include the whole continent up to the Eurasian plate, which meets it at Turkey, unfortunately for the Turks.

The Australian plate includes Australia and India. The Pacific plate includes that ocean and the islands on it. One of the other small plates is the Antarctic, which includes everything south of Australia and the Indian Ocean.

There are a couple of others that cause some problems from time to time.

Under this theory the Atlantic is still growing, while the Pacific is shrinking. The Mid-Atlantic Ridge is pushing the Americas up against the Pacific plate and they are diving down the Pacific Ocean trenches.

Volcanoes and earthquakes are the natural result of the conflict between the plates, as we shall now see.

8
Volcano!

> Nothing which we can imagine about Nature is incredible.
>
> Pliny the Elder

Near the Mexican village of Parícutin there once was a ranch called Tepacua. On the land of Tepacua there was a hole.

The tenant farmers used to dump things in the hole, which had all the appearance of a bottomless pit. No matter how much they dumped into the hole, the hole never seemed to fill.

Early in the year 1943, farmers noticed an occasional rumble underground. No one thought much of it.

On February 20, Dionisio Pulido had been walking through the field with his wife and son. He noticed something odd about that hole—it seemed to be widening.

He felt a sudden thunder. "The trees trembled," he said. He looked at the hole. It was suddenly moving. The ground swelled and raised itself.

A crack formed and dust or ash poured out of the crack.

"Immediately more smoke began to rise with a hiss or whistle, loud and continuous, and there was the smell of sulfur. I then became greatly frightened and tried to help unyoke one of the ox teams.

"I hardly knew what to do, so stunned was I before this, not knowing what to think or what to do and not able to find my wife or my son or my animals."

He ran past a stream and noticed there was no longer water in it. He got on his horse and galloped into town, where he reported what had happened and, later, located his family.

At about the same time a farmer's wife, Aurora de Cuara, felt tremors and noticed a crack open up in the ground. Sparks rose up from the fissure.

In time, huge stones were being thrown up through the fissure and these were followed by incandescent rocks. Eventually the rocks became fire bombs, hurled through the billowing smoke, forming flashes of lightning in the twisting gloom.

Parícutin was about to see a mountain being born the hard way. It was a brand new volcano.

By eight o'clock the next morning the mound around the fissure had risen to about 30 feet (10 m). It was still billowing smoke and tossing out boulders. Later in the day it had grown to twice that size, and the incandescent material was building up a cone.

Two days later lava began to flow from the cracks, spreading over the field at about 16.5 feet (5 m) an hour.

Six days after the initial eruption the volcano was 525 feet (160 m) tall, and was working like a maddened cannon, hurling blocks of stone surrounded by fire, to the accompaniment of gigantic booms.

A month later it had formed a mountain 1,500 feet (450 m) high. The towering smoke had reached a height of more than 20,000 feet (6,100 m). Ash covered the countryside for miles around. Forests died in flame, and two towns were destroyed.

It took a month to build the mountain of Parícutin. The local residents were very lucky. History is replete with people who were not so lucky.

In 1902 the volcano Pelée on the island of Martinique erupted violently. The suffocating cloud that resulted killed 40,000 people and completely destroyed the largest town on the island, St. Pierre.

The most famous eruption was in A.D. 79 when Mt. Vesuvius, near Rome, blew up, destroying three towns—Pompeii, Hercula-

The classic volcano. This one is Cerro Negro in Nicaragua, born in 1850. There are similar, but now dormant, volcanoes in the American West

neum, and Stabiae. Many of the residents were caught by the eruption, and their bodies were found centuries later as they had fallen, frozen in time and horror.

Pliny the Elder, a Roman scientist who thought there was nothing incredible about nature, sailed to the scene with a fleet. He was caught in a cloud of poison gas and died—neither the first nor the last victim of air pollution.

In 1883 the volcano Krakatoa in Indonesia exploded and killed about 36,000 people. There are scientists who believe that it may have produced the loudest noise in the history of the world. Some of the explosions were heard in India and Australia, 3,000 miles (5,000 km) away. The dust collected in the upper atmosphere and colored sunsets around the world for months.

The volcano Santorini may have brought the end of the ancient Minoan civilization on the island of Crete.

Mt. Etna on the island of Sicily erupts periodically, chasing the nearby residents from their homes and destroying their farms and villages. For some unaccountable reason they move back as soon as the ash clears, and so are in the right position to be wiped out again by the next eruption.

There are several active and spectacular volcanoes on the Hawaiian Islands, which are, to be perfectly correct, volcanic mountains sticking up through the water.

In 1970, the Icelandic volcano Hekla began an eruption that lasted two months. Within two hours of the eruption something like 230 million cubic feet (6.5 million m^3) of volcanic ash had been thrown thousands of feet into the air. The gas burping violently out of Hekla is heavy with fluorine, which is poisonous. That is the same stuff that helped kill all those people in Martinique.

It was a spectacular eruption, but Iceland has seen those before. In 1783 the crater Laki erupted, and the resultant damage killed 11,500 cattle, 28,000 horses, and 190,500 sheep, about 50 percent of all the livestock on the island. This caused a famine that killed off 9,500 people, about 20 percent of the Icelandic population.

Iceland, you remember, is a part of the Mid-Atlantic Ridge sticking up out of the water.

Volcanic explosions sometimes trigger tidal waves, adding to the chaos.

Some volcanoes are just plain mean, erupting time and time again.

Vesuvius, after the catastrophe of A.D. 79, had occasional eruptions until 1139, then began a period of new activity in 1631, with extensive eruptions in 1766, 1779, 1794, 1822, 1872, 1906, 1929, and 1944. In the 1631 eruption five towns were destroyed and 4,000 people were killed.

Krakatoa erupted in 1927, the eruption forming an island, Anak Krakatoa. That island exploded two years later, forming still another island in the hole created by the 1883 blast.

Pelée, which wiped out much of Martinique, erupted once more on September 16, 1929, and kept blowing for three years.

The island of Bali is particularly plagued. Mt. Agung (10,308 feet high or 3,142 m) erupted three times in 1963, killing 1,500 people and destroying about a third of the island's farmland. Mt. Batur (5,636 ft. or 1,718 m) erupted later the same year, forcing 1,200 people from their homes.

Volcanoes come in large sizes. Guallatiri, which last erupted in 1959, is 19,882 feet tall (6,060 m). Ernest Hemingway's fabled Kilimanjaro, which is dormant, is 19,340 feet tall (5,895 m). There are thirteen volcanoes more than 10,000 feet (3,048 m) tall in South America alone. There are nine in North America, topped by Citlaltépetl (18,700 ft., or 5,700 m) and Popocatépetl (17,887 ft. or 5,452 m).

Wrangell in Alaska is the tallest in the United States (14,163 ft., or 4,317 m). Alaska is loaded with volcanoes. So, of course, is Hawaii.

The world's champion self-destructing volcano is probably Bandai-san in northern Japan, which blew itself up in 1888. It is believed to have blasted 1,587,000,000 cubic yards (1,213,020,000 m^3) of material over a 27 square mile (70 km^2) area.

The stuff that causes the violence is called magma, which is molten rock charged with water vapor or gases under pressure —like a pressure cooker. The magma is located about 25 to 30 miles (40–48 km) below the surface.

Depending upon circumstances, there are several kinds of

volcanic eruptions.

The Hawaiian type are the most peaceful—usually a lot of lava bubbling to the surface, forming unbelievably beautiful fountains of fire.

Next up the scale of violence is the Strombolian type, named for the volcano off the Sicilian coast. The lava is thicker and there is more gas, so when it blows it does so with considerable gusto. The lava forms balls as it hardens in the cool air and comes back to the ground as fire bombs.

Vulcanian eruptions contain magma so thick it forms boulders while it is still rising through the fissure. The result is that it blasts these blocks of rock like cannonballs through the air. There is usually a considerable amount of ash and gas.

In the Vesuvian volcano, the pressure is relieved while the magma is still going up the tube through the crust. This produces a gigantic column of gas blowing out the top, and bores holes through the cone. Lava flows from new fissures.

Plinian eruptions are named for Pliny. Melted rock is torn into pieces, and in many Plinian eruptions the explosion is so violent it blows the top off the cone. The most violent eruption in recorded history was in Tambora, Indonesia, in 1815. It blew the top off the mountain, which began the day with a height of 13,000 feet (3,960 m) and ended the day down to 9,000 feet (2,750 m). Vesuvius once lost 1,600 feet (485 m) in such an explosion.

The worst volcanic eruption, the pelean, almost defies description. It was named after the Martinique mountain and it produces what is called a *nuée ardente*, a cloud of fire.

This kind of eruption forms lava bubbles that, when they burst, bowl down the side of the cone at more than 150 miles an hour (241 kph). Every single particle in the eruption is giving off a gas simultaneously, while clashing with others and forming a cloud that destroys everything in its path.

One of the things in the path of the 1902 Martinique explosion was St. Pierre, which simply did not exist once the *nuée ardente* had passed by.

Some volcanoes have left behind vast areas of cold lava, which covers the crust like armor plating. These volcanoes are in the form of huge fissures through which the molten material pours.

Out of a sea of lava, an eruption. This is one of the many active volcanoes in Hawaii

Parts of California, Idaho, Oregon, and Washington State are covered with just such a lava flow. The sheets covering Iceland may be the largest, spreading about 300,000 square miles (777,000 km^2).

Many volcanoes are under water and the fact that they are submerged does not prevent them from blowing their tops, with sometimes spectacular results. Etna and Vesuvius started life that way.

Most volcanoes are cone-shaped. A minority are in the form of flat domes.

The mounds are generally stratified in nature, layers of cold lava intermixed with layers of rock and rubble. These are called composite cones, and the layers are due to repeated eruptions.

Sometimes after a particularly violent explosion (Plinian, usually) a large bowl, called a caldera, will form in the center of the mound. Frequently the calderas fill with water and form crater lakes.

Volcanoes are always associated with two things—a weakness in the Earth's crust that will permit the rising lava to blow through, and a supply of magma to do the blowing.

Once the magma gives out, the volcano stops erupting and goes into a state in which it puffs out gases, called the fumarolic stage. This is followed by a period of hot springs, then cold springs, and finally the volcano becomes completely dormant. Erosion wears it down until finally all that is left is the original tube of lava, the volcanic pipes found by the explorers of the *Alvin.* The diamond mines of South Africa are the relics of such old tubes.

Most volcanoes occur in bands, usually along mountain ranges like the Andes and wherever there is extensive geologic activity. There is a ring of volcanoes around the Pacific, the so-called "Ring of Fire."

You almost always find volcanoes over descending plates, where one continental plate dives under another, such as the west coast of South America and Alaska.

Where there are volcanoes, there are almost always earthquakes. Volcanoes may, in fact, cause some earthquakes. The heat from the magma produces a giant belch in the crust.

How to take a volcano's temperature. An American geologist films the activity of La Soufrière on the island of St. Vincent in the West Indies

A fireball of lava boils up out of a fiery ocean of molten material. It sinks only to rise again. This is Kilauea, Hawaii

This occurs in a place called the Benioff Zone, an area that is also the site of deep quakes. There is apparently great heat in the zone, and this may be enough to weaken the crust and let the pressure escape.

Part of the pressure comes from the presence of water in the magma. It is something like the reaction you get when you put a hot, greasy pan under cold water.

The water in volcanoes is generally in the form of steam. The steam pressure is great enough to blow the rocks that plug up volcanic pipes, and the volcano erupts.

This vapor, incidentally, is also the source of atmospheric moisture and surface water. The gases in volcanoes provided us with our atmosphere.

If you really want to see volcanic action you should go, not surprisingly, to Iceland.

The Icelanders are subject to a particularly horrifying form of volcanism called *jökulhlaup*. Many a mountain in Iceland is snow-capped (a *jökull*). Unfortunately some of these *jökulls* are also volcanoes.

The heat from the volcanoes melts the snow on top of the mountain until there is not enough snow left to hold back the eruption. All that melted snow is now in the form of water, so when a *jökull* erupts, a *jökulhlaup*, the greatest danger is from water, not fire. It is like a dam collapsing.

Since Iceland is just an exposed part of the Mid-Atlantic Ridge, it is safe to say that the forms of volcanic activity in Iceland are the same as those found in the ocean—not the snow-capped mountains, of course.

Icelandic volcanoes tend to be slashes, not circular in shape. The island is scoured with rifts and troughs.

Sitting at the end of one of the rifts is Hekla, hell in Icelandic folklore. Hekla is the volcano that erupted in 1970.

South of Hekla are a bunch of volcanic islands, one of which is called Heimaey. On Heimaey was the fishing village of Vestmannaeyjar. It seemed a safe place to be because Helgafell volcano there had been dormant for about 6,000 years.

The first inkling that something really spectacular was going to happen was in 1963 when a fishing boat noticed that a section of the ocean was beginning to boil. The boiling was replaced with a full-blown eruption and the result was that the world had a new island. It was called Surtsey. Two more small islands were formed before the volcano went to sleep.

The sleep lasted until January 21, 1973. Then there were some seismic rumblings, and they were repeated the next day. On January 23 a fissure opened on the old volcano, and before long fountains of fire were spurting up out of this volcano that had not been awake for 6,000 years.

Helgafell bombarded the town with rocks and fire bombs. Lava set fire to the grass. The town was being destroyed.

With some help from the U.S. Geological Survey, the Icelanders tried to cool Helgafell down before everything was destroyed, or to divert the lava. They even planned to put it out by blasting away enough rock to let the seawater pour into the fissure. It eventually occurred to them that this might just produce an explosion—something in the order of an atomic bomb—so the idea was quickly dropped.

After a few months Helgafell subsided again, and people returned to what was left of Vestmannaeyjar.

The role of volcanoes in building our Earth should not be underestimated. It is a main force.

One form of volcanism forces the mid-ocean rift to spread. We showed earlier how the lava pushes its way up through the sea floor from the mantle, pushing the floor out of its way in both directions.

Scientists have found two kinds of materials at the rift. One is called volcanic, the other oceanic. The volcanic is on top. It consists of material called tholeiite, rich in aluminum. The oceanic layer is believed to come from the mantle as well, but no one knows very much about it.

Scientists do believe that the flow of volcanic material through the rift is fairly constant. They have been able to measure this, because the faster the sea floor spreads, the less time there is for the volcanic layer to accumulate, and the thinner it is. The volcanic layer can be several miles thick in slow-moving parts of the ocean.

The volcanic flow has been measured at about 4 cubic

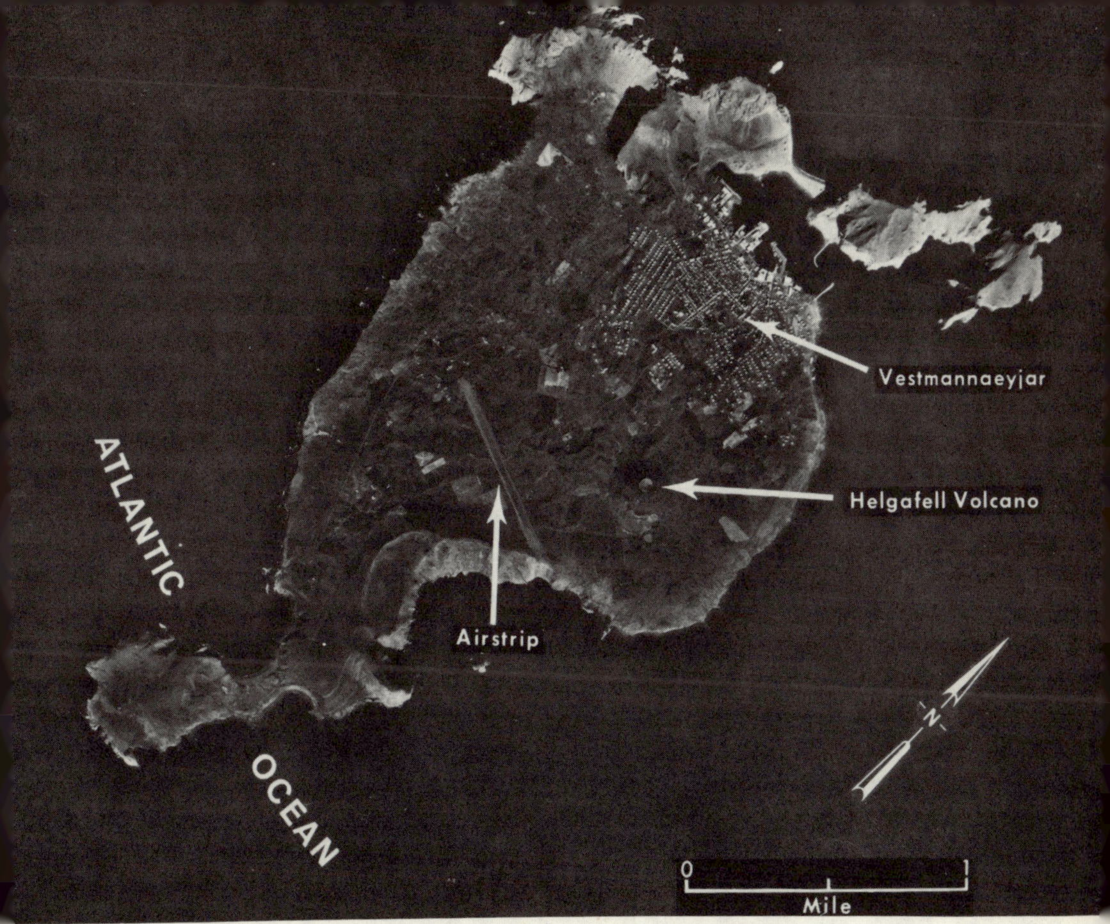

GS

The island of Heimaey and the Helgafell volcano. This is an aerial photo made by the U.S. Geological Survey which went to Iceland to help out in 1973

kilometers per year (140,000 cu. ft. or so) in all the rifts. That is much more than the volcanic action on land.

Since you almost never know about volcanic action underwater, it is clear that what we do see on the continents is only a fraction of what is bubbling away beneath the waves.

What oozes out from the rift is only part of it. As the sea floor moves, more lava erupts through volcanoes and flows, building the mountains that form this gigantic mountain chain.

That's why things are so busy in Iceland, a part of the chain that sticks out of the water.

Iceland is cut almost in half by a rift that runs parallel to the mountain ridges. The rift is almost 45 kilometers wide at one end (28 mi.). Lava flows through fissures at the rift, forming long, low volcanoes, many of them active. According to H. W. Menard, the volcanoes are formed at the rate of one every five years, and more than two dozen are currently active.

The speed of spreading also affects the size and shape of these volcanic mountains—the slower the spreading, the higher the mountains.

In places where the spreading is only .8 to 1.4 inches per year (2–3.5 cm), the volcanic mountains are quite high. Where the spreading is between 2 and 4.75 inches (5–12 cm) the mountains are usually long and under 1,640 feet (500 m) high.

The farther away you get from the rift, the less volcanism you find, until you get to the trenches. Hawaii, which consists of round volcanoes, is smack in the middle of the Pacific plate.

What seems to slow down the activity is the aging of the plate. As the lava hardens and moves, it seals some of the dikes that channeled the lava from the interior. But it never completely seals.

Even crust 10 million years old carries large, active volcanoes. In fact, says Menard, it takes about 10 million years for a volcano to get big enough to stick its head up through the waves and become an island.

There are some volcanoes sitting on crust 100 million years old—the Canary Islands. That is unusual, Menard says, because usually volcanoes sputter out after 20 to 30 million years.

Also, some volcanoes never survive the move. If they are not active enough to produce 100 cubic kilometers of lava (about 3,500,000 cu. ft.) per million years, they will not make it to the island stage. The moving plate will sink them before they get a chance to nose through the surface.

Guyots, mentioned earlier, are the remains of volcanoes that didn't make it. Most are found on crust 30 million years old or older.

One problem confronting scientists was the concept of a volcano sliding on the crust while still tooting up lava from the mantle. The volcanoes travel hundreds of miles, and to accept this notion you have to accept the fact that there are magma chambers big enough to supply these moving chimneys.

"It remains perfectly possible for a volcano to drift for tens of millions of years over hundreds of kilometers [or miles] while tapping a single magma chamber embedded deep in the mantle," Menard wrote in *Scientific American.*

He also pointed out an alternative theory, which seems to him to be equally valid—that the volcano and its tube to the mantle travel along with the moving crust, tapping different parts of a stationary magma source. In other words, there may be no real, contained magma chambers as such, but simply an area widely filled with the hot stuff.

There are approximately three to four hundred active volcanoes in the world.

So you see that volcanoes have played a vital part in the construction of our world, and in the history of man.

Having finished this chapter, you might want to read one of the oldest existing descriptions of the kind of destruction volcanoes can bring.

Just south of what is now the Dead Sea, a section in which the

A little quieter, Mt. Erebus in the Antarctic, still sends up smoke and gasses. Japanese scientists crawl down into it periodically for measurements—very dangerous work

mid-ocean ridge crosses onto the continental land mass, there used to be two cities, Sodom and Gomorrah.

There is nothing left of Gomorrah. At the traditional site of Sodom there is now an Israeli potash mine and the dead, salt sea.

Once, it is believed, thousands of people lived there in unrepenting evil.

"Then the Lord rained on Sodom and Gomorrah brimstone and fire from the Lord out of heaven; and he overthrew those cities, and all the valley, and all the inhabitants of the cities, and what grew on the ground. . . . And Abraham went early in the morning to the place where he had stood before the Lord; and he looked down toward Sodom and Gomorrah and toward all the land of the valley, and beheld, and lo, the smoke of the land went up like the smoke of a furnace."

You can find this description in the Bible: Genesis, chapter 19!

9
Earthquake!

> I fled, and cry'd out, death!
> Hell trembled at the hideous name, and sigh'd
> From all her caves, and back resounded, Death!
>
> John Milton

First the sea gulls rise, circle overhead, and fly quickly out to sea.

Horses, skittish for hours, suddenly panic and begin to scream.

Dogs and cats begin to howl. Rats flee from cellars.

And then the Earth begins to moan.

It is a low sound, a sigh. It moves closer and closer and closer still, like the sound of a giant cymbal or an approaching train.

Louder. Closer. The Earth shakes. Earthquake!

It is the most destructive force in nature. Millions have died through history. Entire cities have been wiped out, whole regions devastated. Over twenty thousand people died in quakes in Guatemala alone in February of 1976.

And yet they are inevitable. On every living planet made of rocks and stones, there has to be an earthquake at one time or another.

Scientists are just now beginning to understand earthquakes and how they come about. They are on the verge of being able to predict them in advance, and someday in the future they may even be able to prevent them.

Anchorage, Alaska, in 1964, after a major earthquake. Notice the cracks and shift in the ground. The rubble in the upper right of the picture was a six-story apartment house under construction

SGS

Charleston, S.C., after the quake of 1886

How serious is the problem?

On April 18, 1906, an earthquake struck the beautiful city of San Francisco. When it was done, and when the fires it caused had subsided, five hundred people were dead and most of the city was a charred ruin.

The former U.S. Office of Emergency Preparedness estimated that a quake that size would leave 10,360 dead and 40,360 hospitalized if it happened now, say, during a rush hour. About the only thing that could happen to San Francisco that would be worse would be an atom bomb.

San Francisco is not the only place in North America where earthquakes have struck. Few places are immune. There has even been an earthquake in New York City, but it was about a hundred years ago, when the city was much smaller and had no skyscrapers. Imagine what would happen if a quake struck there today!

The San Francisco quake is the most famous in American history, but the most powerful was probably the series in the Missouri Valley in 1811–1812. Almost no one lived there at the time so few, if any, people were hurt. But Charleston, South Carolina, was struck in 1886 and sixty people were killed.

The most likely place to experience an earthquake in North America is the Canadian Province of Quebec, which has a strong earthquake every thirty-five years or so.

Some places are more prone than others because most earthquakes seem to be associated with plate tectonics. Many earthquake faults or slippage cracks are located where two plates meet and rub together.

The famous San Andreas fault in California, for instance, is at the juncture of the Pacific and the North American plates. There is that famous "Ring of Fire," the string of volcanoes that runs up and around the Pacific, which is also the site of considerable earthquake activity. It is directly associated with the faults of these gigantic plates.

The San Andreas is the fault that causes most of the damage in California. It was the San Andreas that ripped in 1906, destroying the city of San Francisco. It passes from northern California, just offshore at San Francisco, down east of Los

USGS

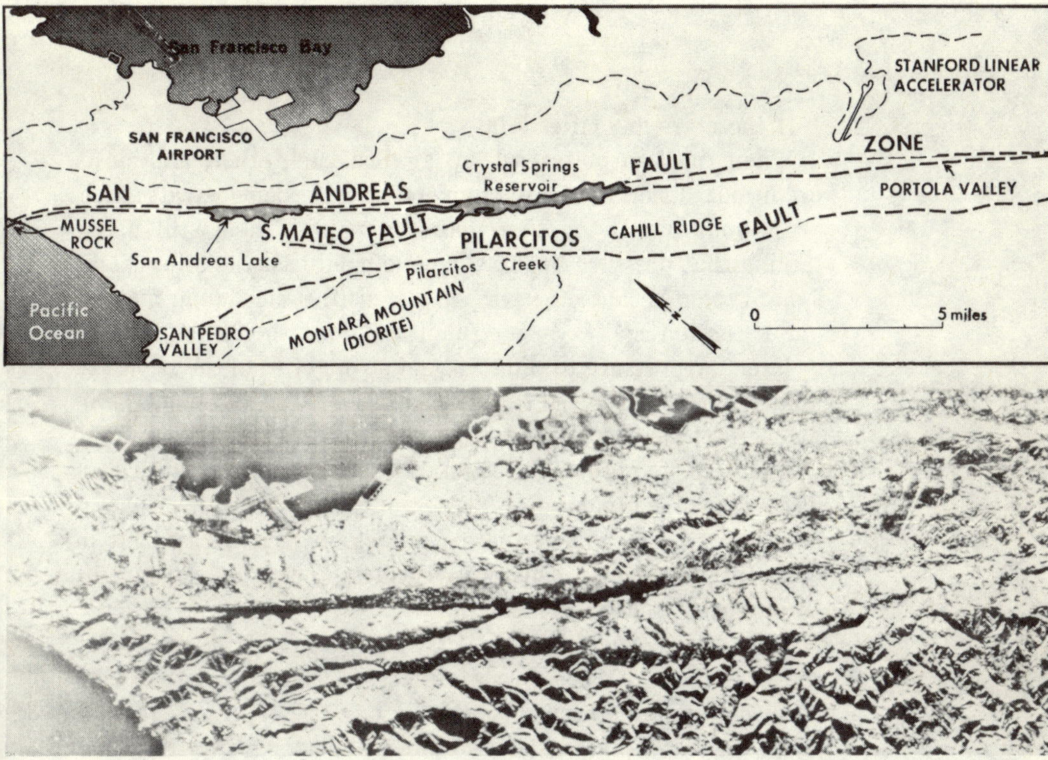

The famous San Andreas fault near San Francisco. The upper picture shows the location of the fault on a conventional map. Below is a radar image

Angeles and on south into Mexico.

Everything west of the fault, which includes the city of Los Angeles, is on the Pacific plate. Everything east of the fault, which includes San Francisco and the rest of the continent, is on the North American plate.

Unfortunately for California, the plates are moving in different directions and rubbing up against each other along the fault. The Pacific plate is moving northwest at about two or three inches a year, heading for Alaska, where it dives under a plate there.

The North American plate is moving southeast. Believe it or not, someday several million years from now Los Angeles will slide right by San Francisco as the plates take the cities in opposite directions.

If there are any cities left.

When the two plates rub up against each other, friction is produced. Friction resists movement. To demonstrate that, press your hands together, palm to palm. Then, still maintaining that pressure, try to slide one hand past the other. This is hard to do because the friction is resisting the sliding motion.

But tension is always being built up along the faults. Sooner or later the pressure to slide will get stronger than the friction's ability to resist, and the two sides of the fault will violently slip by each other. That is an earthquake.

Scientists are pretty sure this is what happens because of a discovery at a famous California winery in Hollister. Several buildings at the winery were found to be moving, pulled in different directions. This was because the fault runs right through the winery.

This motion, called fault creep, was measured and studied for the first time in 1956. According to one of the scientists who studied Hollister, the creep varies along the fault line in that area. In some places the two sides slip along smoothly and peacefully. In other places they jerk and skip as the sides skid against each other, causing frequent minor earth tremors.

In places where the creep is steady and smooth, scientists think major earthquakes are not likely to happen. But in those places along the fault where there is no creep at all—where friction has locked both sides in place—there is great danger. The fault has been locked in San Francisco since the last quake in 1906. It has been locked under Los Angeles since 1857.

The pressures are still there—pressures strong enough to move whole continents.

Dr. Robert Wesson of the U.S. Geological Survey earthquake center in Menlo Park, California, thinks the pressure will never end.

"For ten million years the Pacific has been moving up and under Alaska," he says. "It is not likely to be stopped now."

Sooner or later the friction will no longer be able to hold. The fault will rip violently. The result could be catastrophic. This danger is very real.

"All of us agree San Francisco will be affected by a very

Where the ground opened during the great earthquake of 1906, when the San Andreas shifted. It will happen again

serious earthquake at some time," says Dr. James C. Savage of the National Center for Earthquake Research at Menlo Park.

The danger is equally great in the Oakland-Berkeley area, Los Angeles, San Bernadino, California, and in southern Alaska.

The danger is also one of the reasons research into detecting quakes (and perhaps even preventing the big ones) is one of the most active fields in all of science.

Scientists are following many paths in their earthquake studies.

The Menlo Park facility of the Geological Survey, which is where Dr. Savage and Dr. Wesson work, is typical of the centers around the world.

Into the center pour data from an extensive network of sensors that measure every little movement beneath the Earth's surface. Each ripple and burp appears as a squiggle in a line drawn by robot pens on a roll of paper in a seismograph machine.

Because there are so many sensors (the entire northern end of the San Andreas is wired into the center), scientists are able to pinpoint each quake.

They look at two things. One is the focus, the actual point where the quake occurred beneath the surface. The other is the epicenter, the spot on the surface directly above the focus.

Every quake sends out messages in the form of seismic waves. We measure them on seismic instruments to tell how powerful they are. The most commonly used measurement is the Richter scale, named for famed seismologist Dr. Charles Richter, who, incidentally, lives in Southern California near several seismic faults.

The Richter scale is open-ended. That is, it begins at 0 and goes up infinitely. In fact, the strongest quakes recorded never reach more than 8 or 9 on the Richter scale, although theoretically they can go higher.

The Richter scale measures the amount of energy in the quake by measuring ground vibrations. Scientists can get a Richter reading within moments of a quake. Generally, any quake that gets above 4.5 on the Richter scale will cause damage in a built-up area. More than 6 or 7 and there is a disaster.

Geophysicist David Carver watches the seismogram at the Geological Survey's Boulder, Colorado, earthquake center. It constantly monitors the fault

There are three kinds of seismic waves, and each one acts differently, so each gives its own clue as to what is going on.

Pressure waves, or P waves, move through the earth like sound waves, bulging and contracting. Shear, or S waves, move out at right angles. Surface waves are the third kind, but they are felt only within about the top 20 miles (32 km) of the Earth's crust.

P waves and S waves travel at different speeds. S waves are about one third slower.

Russian and American scientists have used this speed difference as a major clue in predicting quakes. They have noticed that in periods before big earthquakes, when there were lots of tiny tremors, the difference in speed between the S waves and the P waves varied strangely. Something had altered the rocks through which the waves traveled.

They now think that by monitoring these little quakes they can get some hints as to when a big one is getting worked up.

Other scientists are working on the idea that the ground tilts by very tiny amounts before an earthquake. They know it happens on the side of a volcano before an eruption. Japanese scientists are now pretty sure the same thing happens before an earthquake occurs.

Others believe the magnetic force or electrical conductivity in rocks changes before a quake. Some others think they can watch to see where quakes occur along a fault line and determine areas where a big quake can be expected.

Maybe the underground water table gives us clues.

If we know what causes quakes, can we ever prevent them from happening?

Well, there are a few scientists who think we can. They think so because of a very odd thing that happened in Colorado about ten years ago.

There is a lovely word in our language, "serendipity." It is serendipity when you unexpectedly stumble upon something marvelous. Serendipity is what happened in a very unlikely way in Colorado.

Near the Denver airport is a United States Army depot for poison gas and chemicals. For years the Army was pumping some of the old chemicals underground to get rid of them.

Denver suddenly was becoming the scene of some moderate-sized earthquakes. This was unusual, because Denver was not near any fault known then, and no quakes of any substantial size had ever been recorded there before.

Someone began to keep a record of the quakes and related them to the chemical dumps. He found that whenever the liquids were being injected underground, Denver had an earthquake.

A similar thing was happening nearby, in an oil field near the

USGS

Olive View Hospital in suburban Los Angeles began life as a six-story building. After the San Fernando quake in 1971 it was only five stories and had to be abandoned

little town of Rangely, Colorado. Whenever the oil company pumped water into a well to force the oil up, there was a quake in the field.

In 1968 scientists moved into the Rangely field and borrowed four wells.

They demonstrated that whenever they pumped in water they produced quakes. Whenever they stopped pumping, the quakes stopped.

It now seems that there is a fault under the Denver area. The liquid being dumped underground finds its way into the fault, separates the two sides, and cuts down the friction, allowing the two plates to slide.

Go back to that demonstration of putting your palms together. Now try it with wet, soapy hands. You will see that the friction does not hold and the hands slip right past each other.

The scientists stopped the quakes in Denver by halting the underground dumping and so locking the fault.

The line through the center of this picture is the San Andreas fault. Despite its deadly history people build houses and reservoirs right over the fault

This pawnshop in Anchorage did not float a heavy loan. An earthquake did this

The "Ring of Fire" and the zone of earthquake danger around the world. The "Ring" runs from Manila, around the top of the Pacific, down the American coast, and into Chile

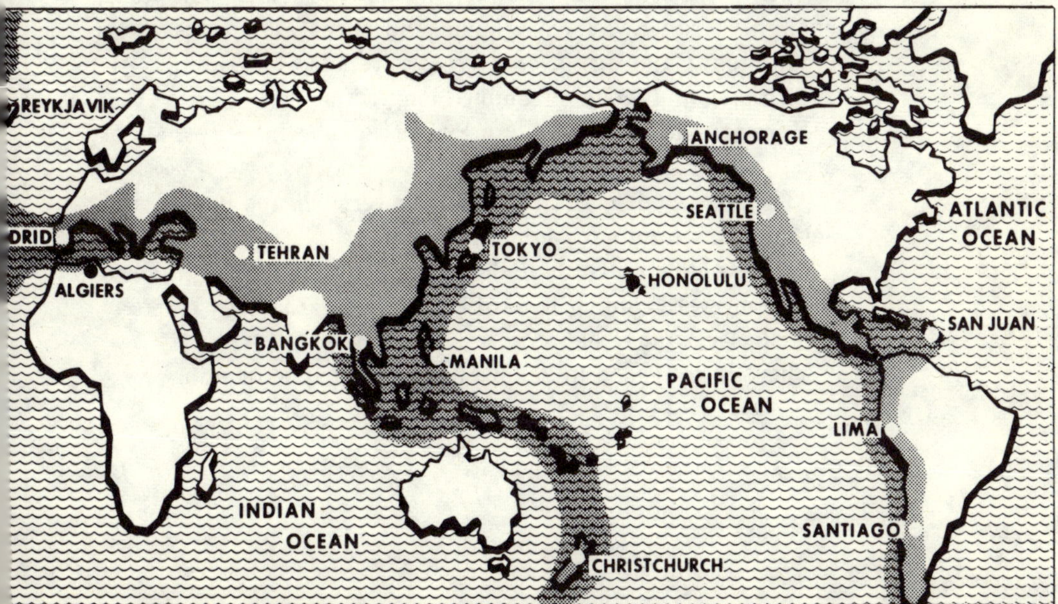

Now go one step further. If you can cause minor quakes by reducing the friction, isn't it possible to prevent major quakes by unlocking the faults? In theory, all you would have to do is inject water into the San Andreas fault, for instance, and accept the moderate quakes that this would cause. For at the same time you would be preventing the kind of devastating quakes that occur when the fault is locked for long periods.

We are a long way from doing that yet.

Not all quakes have anything to do with faults. Most Eastern U.S. and Canadian quakes are not associated with known faults, according to geologists at Penn State University. Predicting them is beyond our capabilities at this time.

The only answer is to protect against the likelihood of damage by building quake-proof buildings like those now required by law in California.

It is not likely you will be hurt by a quake, even in California. The people at Menlo Park, who study them all the time, live in a town right alongside the San Andreas fault, and they and their families do not live in great fear.

But the danger is there, and they and others around the world think they can soon find a reliable way to predict quakes and to save the lives lost when the Earth moves.

Earthquakes are terrible things. After the 1971 earthquake along the San Fernando fault near Los Angeles (a fault most scientists thought was inactive), family dogs were frightened enough to run away from home and form wild packs in the mountains. Children suffered great psychological trauma. And grown men and women trembled long after.

Here are the words of a San Francisco policeman, Jesse Cook, one morning in 1906:

"There was a deep rumble, deep and terrible, and I could see it actually coming up Washington Street.

"The whole street was undulating. It was as if the waves of the ocean were coming towards me, billowing as they came."

Soon Sergeant Cook's city was in rubble.

There is no reason to believe it will not happen again.

10
How to Build a Mountain

> On the mountains there is freedom!
> The world is perfect everywhere,
> Save where man comes with his torment.
>
> Schiller

Mountains.

The passion with which men have loved mountains knows no bounds, and anyone who has spent time among them—the Alps, the Rockies, the Andes, the Transantarctic, The Himalayas—shares that passion, understands it, transmits it to others. Mountains are among the most beautiful things nature can produce.

How nature builds mountains is an integral part of our story. It should be clear by now that the Earth is quite dynamic—shifting, groaning, burping, cracking, tearing. It would be impossible for the crust to move around the way it does without something crumpling, or poking up, or erupting. Essentially, this is exactly how mountains are formed.

There are very few lone mountains. Most mountains dwell together in mountain chains or ranges (scientists prefer the term "cordilleras"). There are some exceptions, usually mountains produced by erosion.

Stone Mountain, near Atlanta, is one of those exceptions. It was once a huge boulder of granite that was buried in a plain.

The plain was eroded away but the boulder, much harder than the surrounding plain, remained uncovered as a mountain.

The geologic term for that is "monadnock," named for another mountain in New Hampshire.

Stone Mountain is isolated. You must drive for hours to get into other mountains. But this is unusual. Usually where you see one mountain, you see lots of mountains.

They look ageless, but of course they are not. It is a general rule (but all rules have exceptions) that the higher the mountain is, the younger it is.

The Appalachians, for instance, once stood as high as the Rockies. But through the eons they have been eroded to the point where they are only ghosts of their former selves, lovely as they may be.

The great mountain chains are much younger than the continents they ride upon. The oldest ones no longer exist, having been scraped down to the point where the land is flat and only the buried stumps remain.

Throughout the history of Earth the building (orogeny) and eroding of mountains has been going on, and still is.

How does nature build mountains? There are several ways. We have seen how erosion produces monadnocks. Erosion can also produce clusters of mountains. You see them in the western American desert and in the Ozarks.

But there are four major geophysical methods of orogeny. The commonest method can be demonstrated with a tablecloth or a towel.

Take the cloth or towel and spread it out flat. Now push two opposite sides toward each other. You will see that the cloth rumples in several places. Sometimes a hump will fall over on its side.

That is what happens to the crust of the Earth, which is always moving. The humps are mountains, and the loops that fall over are called nappes.

Some mountains are the result of faults, those cracks in the crust frequently associated with earthquakes. If one side of a fault is thrust up higher than the other side, it can be called a mountain range. Usually there is more than one fault involved

—there are generally rows of them, all upthrusting at the same time to the accompaniment of quakes and tremors.

Some of this is found in the Alps, Appalachians, and Rockies, demonstrating, among other things, that there is usually more than one cause of giant mountain chains.

Some mountains are the products of "intrusion," which means that something below the surface pushes up, creating bulges. Sometimes it is the continental basement itself, punching up through the top. Sometimes it is an intrusion of molten rock forming a very large dome.

Finally, there are volcanoes.

Anytime you study one of the great cordilleras, you are likely to find evidence of all four kinds of action.

Geologists have long been puzzled by mountains. They could not imagine how they were built. They resorted to such models as a shrinking apple. If you take an apple and let it dry out, the apple will shrink and the skin shrivel. They believed the shriveling of the skin of the Earth's crust was the same thing. They were wrong—the Earth has not shrunk, but the concept of a skin slipping and sliding around was not too far wrong.

There were some mysteries that needed solving. Back in 1450 the Swiss scientist Felix Hemerli noted sea fossils embedded in Alpine rocks thousands of feet above sea level and far from anything resembling a sea. Similar fossils were found in The Himalayas. It was clear that rocks that formed mountains were once part of the sea, but that made no sense to anyone.

Further, the layers of rocks were all mixed up. There were places where older rock could be found *below* younger rock, as if someone had turned the rock upside down.

The concept of nappes, these fallen-over humps, was first worked out by Arnold Escher in the early nineteenth century.

At about the turn of this century engineers began to dig the Simplon Tunnel through the southern Alps. This immense undertaking, 12 miles (19 km) long, made it possible for geologists to see just how enormous these nappes were. They could also be identified as parts of structures that ranged farther to the south. For instance, some of the nappes were traced from northern Italy into the northern Alps. One nappe located at the Italian-Swiss border has been traced all the way down to the Mediterranean!

Another peculiarity was that some of the sediment rocks found in mountain structures were much thicker than the same rock away from the mountains.

At about the time of the Civil War, American James Hall compared rock found in the Appalachians with similar rock in the prairie states. One of the mountain layers was 32,800 feet (10,000 m) thick, while that same layer in the Mississippi Valley was much thinner. There was a clue in there somewhere.

Hall guessed that somehow sediment had built up in the mountain area, that the weight of it had pushed down the crust, and that eventually the crust broke under the strain and the sediment was popped up.

James Dwight Dana, in 1873, however, decided that could not be the case. The sediment could not possibly weigh enough to cause that kind of rumpus. He went to the shrunken apple theory and said that the shrinking, caused by the continued cooling of the Earth's interior, caused troughs. Sediment gathered in these troughs, which somehow pushed up mountains parallel to the troughs. He named the troughs "geosynclines," a term used by geologists now to describe mountain-building troughs, but the present definition is not the same as Dana's.

Then came Wegener. A scientist named Emile Argand figured that if Wegener was right, the plates crushing against each other could rumple the crust into mountains.

Now if you combine the two ideas, continental drift and geosynclines, you have the modern explanation for mountains. The upthrusting is considerable. Someone once remarked that if you flattened out Switzerland it would be gigantic. Geologist Rudolf Staub was more scientific.

"If we smooth out only the Alpine folds and sheets over the traverse section between the Black Forest [in Germany] and Africa then in relation to the present-day distances of about 1,800 km [1,100 miles], the original distance separating the two must have been about 3,000 to 3,500 km [1,800–2,100 miles], which means an alpine compression . . . of around 1,500 km [900 miles]."

That means the distance between the Black Forest and Africa was once about 1,800–2,100 miles and it has been squeezed almost in half, the shrunken land turned into mountains.

Another clue that led scientists to the present theory is that most geosynclines appear to be either on the edges of continents or in between two of them.

James Hutton had said the formation of the Earth was the result of slow, continuing processes, so if there were geosynclines then, there must be geosynclines now building mountains.

Well, scientists couldn't find them. They were looking in the wrong place.

There is at least one, though it is underwater off the eastern coast of the United States. It is just like the geosyncline that built the Appalachians.

In fact, studying how the Appalachians were probably built is an excellent way of studying how mountains are built. The Appalachians were once very grand, and their grandeur can be traced back to the birth, death, and rebirth of the Atlantic Ocean.

The best description of this may come from Robert Dietz of the National Oceanic and Atmospheric Administration (NOAA), writing in *Scientific American*:

Remember those continental plates? Assume for the sake of the discussion that the plates are rectangles (only the Indian plate comes close to that shape in fact).

The rectangular plate has four sides. One side is the side that comes pouring out of the ocean rift. One side is the side that dives into a trench. The other two rub up against other plates.

Sitting on top of those plates are the continents, which geologists call cratons.

The edge of the continent farthest from the rift, the one moving toward the trench, can be called the leading edge. The back of the continent, the one closest to the rift, can be called the trailing edge, because it is trailing behind the other.

The trailing edge of the North American plate is quite stable. The Continental Divide—that line between the Atlantic and Pacific drainage areas—is much closer to the West Coast, the leading edge, so most of the sediment washed down by the continent's rivers heads toward the east, the trailing edge. The sediment builds up east of that continental rim in the ocean. There, it forms a geosyncline, the continental slope.

How did such a geosyncline build the Appalachians?

Remember, there were two Atlantics; the sea opened and closed, then opened again. Now consider a geosyncline being built on the East Coast of the North American plate during the first Atlantic. The Atlantic continued to spread. As the molten material poured out the rift, the area around the rift in the middle of the ocean rose in a sort of bulge.

Eventually the part of the plate at the geosyncline began to dive under this bulge, crumbling the sediment and pushing it up. Molten granite in the plate burst up through the uplifted rocks, adding volcanic action and upthrusting to the crumple.

Then the ocean began to close and the two shores came toward each other. The uplifted geosyncline on the East Coast met violently with the sediment built up on the other side (the west coast of Europe), maybe crushing it if it had not already been crumpled.

Then the ocean started to open again. The uplifted material that had been attached to the trailing edges of both coasts split in two—perhaps not at the same seam, which would account for the great similarities between some North American mountains and some found in Europe. As the continents parted, the mountains sailed away from each other.

Eventually a coastal plain was built up behind the mountains, and the American uplifted area became the Appalachians.

Going from east to west, the first line of rumples was the Piedmont Region, behind it were Blue Ridge Mountains, the ridge and valley belt, the Alleghenies, and the flattened plateau.

The best way to see this is to fly in a small plane over central Pennsylvania near Harrisburg. The mountain-valley complex looks just as if someone had rumpled a heavy tablecloth.

The fact that there is another geosyncline abuilding offshore means that should the Atlantic close again, another line of mountains could be formed in the Eastern United States.

As the land masses and continents clanged and banged into each other, bodies of water were opened wider, creating

Central Pennsylvania and the Susquehanna River Valley. The rumpling of the mountains is obvious. The river running up the center and left is the Susquehanna. The river to the right is the Delaware. Across the right is the Schuylkill

troughs. Into the troughs poured sediment. The land masses kept on moving, plowing through the sediment-filled troughs, producing mountains.

Once Europe was connected to Africa. Eventually they began to split apart, twisting apart the land bridge between the two. Pieces of that land bridge slid around. One became the Iberian Peninsula (Spain and Portugal). It opened the Bay of Biscay and a geosyncline in it.

Later the two continents were pushed back together again. The Iberian Peninsula was jammed up against Europe, scraping through the geosyncline it had created and forming the Pyrenees.

Africa and Europe parted. A geosyncline was formed in that body of water produced by the separation, the Tethys Sea. When the two plates reversed course and moved toward each other, the African plate piled up the sediment in the Tethys and produced the Alps. Since the sediment had been underwater, a lot of marine fossils went with it, as Hemerli had discovered.

India jammed into Asia, creating The Himalayas.

In general terms, that's how you build mountains. But that is simplifying things a bit. We have some pretty detailed information about the history of the Appalachians, showing that most mountain chains are quite complex.

There were actually three different periods of upthrusting that produced the Appalachians.

The land on which the Appalachians were built carried the remains of an earlier mountain upheaval, called the Grenville episode. All we have left of those earlier mountains are the Adirondacks of New York State, which were fortunate enough to be in the right place when the Appalachians were built and so were reborn.

Sediment from the Grenville mountains slid into geosynclines that eventually became the Great Smoky Mountains.

About 450 to 420 million years ago the first upthrust of the Appalachians occurred. The thrusting helped form a huge inland sea, running from Vermont and Minnesota to Alabama and Texas. This first upheaval, called the Taconic after the mountains of the same name, also produced the Green Moun-

The Andes Mountains from a satellite photo. The rumpled effect is quite clear

tains of Vermont, the mountains along the Hudson, through New Jersey, the Delaware Water Gap, and the Blue Ridges that begin in Pennsylvania and run into Virginia.

Volcanic islands sprang up through what is now Connecticut and out in the Atlantic.

There was a pause for sediment-making, which filled more geosynclines. Then, there was another uplift which created eastern New England, piling layers of sediment over the volcanic mounds put there earlier. This is called the Acadian episode.

Then came another pause, this time to permit lush jungles to grow, die, and to become coal. The last period, the Alleghenian, produced the Alleghenies of western Pennsylvania, the Poconos in northeastern Pennsylvania, and the Cumberlands in Tennessee.

Now, going back to that Atlantic opening-closing-opening sequence, we will see how it fits with the three orogenic episodes that created the Appalachians.

The Atlantic was fully grown. Sediment built up. The Atlantic started to close. This led to the Taconic episode.

The Atlantic continued to shrink, and Europe and North America were on a collision course. Sediment built up on the East Coast was squeezed up and over the land in the form of nappes—the Acadian period.

The two continents finally slammed together, first in New England, then Maryland. The collision forced up more material as the two continents piled up against each other—the Alleghenian upheaval.

Finally the course was reversed and the two continents split, aided by rift valleys like the one in Africa. One of them was in New Jersey and is called the Newark Trough.

While all this upheaval was going on in America, similar things were happening in Europe, forming the Caledonian Mountains and other mountains of western Europe.

How about the Rockies?

The Rocky Mountains are part of a cordillera that runs from Alaska to the tip of South America and Tierra del Fuego.

During the days of the dinosaurs there was an upheaval

caused by plate tectonics and volcanism that formed part of the Sierra Nevadas and the Klamath Mountains in the Pacific Northwest. The upthrust sent huge slabs of rock sliding eastward. Volcanic activity broke out all along the line of the Rockies.

Finally, where there had been modest mountains, there grew giants. This was probably caused by an upthrust of material below the surface, lifting huge blocks of rock upward. Some of them are still going up.

Incidentally, one indirect result of all this upthrusting is that rivers began to cut deep into the ground as the ground rose. One such river was the Colorado, and the result of that action is the Grand Canyon.

The Sierras were abuilding the same way, with upthrusting and volcanic activity. The California Coastal Range, such as the Verdugo Mountains near Los Angeles, is the crumpled evidence of a collision between the North American plate and an oceanic plate.

How did the mountains find the sediment for a geosyncline? Scientists think it is possible the West Coast of America ran through Idaho and Arizona.

Where did the volcanoes come from? They were produced by the collision of two plates, the North American, and a plate that no longer exists, called the Farallon plate.

So it all ties together. The moving, restless crust of this seemingly restful planet churns, crumples, piles, and tears.

That's how you build a mountain!

11
And Time and the World

"The time has come," the Walrus said,
  "To talk of many things:
Of shoes—and ships—and sealing-wax—
  Of cabbages—and kings—
And why the sea is boiling hot—
  And whether pigs have wings."

Lewis Carroll

We have seen how the Earth was born and some of the things that happened to it. And we have seen some of the things that are still going on. It is time to put it all together, to tell the story of this planet in its proper context—time.

We will see two things: how slowly, methodically, and surely this planet evolved to what it is, and what insignificant creatures we are in its history.

We have said man is but a blink in the history of Earth. That is really an overstatement. Our story on this planet is not as important as a blink. Perhaps a twitch?

The world was formed 4.5 billion years ago. The Iron Catastrophe turned it into a sectioned ball—the atmosphere, the land, and the oceans formed. This all took about a billion years. (We said geophysicists throw around years like confetti.)

After that first billion years the guesswork decreases a bit. We have some records. The oldest rocks are 3.7 billion years old, mostly sedimentary and volcanic in nature. There are not many of these.

The next block of rocks are from an era called the Archean.

They are 2.2 to 2.8 billion years old. Although the records in these rocks are sketchy, it appears that some of the tectonic systems we have described did not quite apply during the Archean period.

For one thing it seems that mountains did not form at the edge of plates as they do now. They seem to have been rumples caused by a thin lithosphere being bubbled and bounced by a very active mantle.

The Archean rocks were basalts and a form of volcanic rock called andesites, rich with iron and magnesium. These rocks weathered and eroded and formed sandstone.

Yet the Earth probably looked on the surface much as it does today. The same surface processes were taking place. There were wind and rain, and things breaking up and pushing, much as there are now.

Sometime about 3.4 billion years ago, before this Archean period, something quite marvelous happened. Life came to this planet.

We are not entirely sure how, but we have a pretty good idea. The exciting thing is that if this idea is right, it is possible this accident called life may have happened elsewhere in the universe—maybe even before it happened on Earth.

We must now introduce you to an amazing concoction called primordial soup.

It is believed the early atmosphere that burped from the rocks was made up of hydrogen, methane, ammonia, and water vapor.

Methane is a foul-smelling gas sometimes called swamp or marsh gas. It burns and is sometimes used as a fuel. Ammonia is just what you think it is, the stuff in cleaners.

There was probably no oxygen in the atmosphere except that in the water vapor. Oxygen is a product of life.

There was nothing in the gases to protect this atmosphere, or the oceans below it, from the ultraviolet radiation that is generated by the sun.

There were, in addition, great storms producing huge bolts of lightning that zapped through the atmosphere.

This concoction of chemicals and energy, this primordial soup, we think produced life. Something like this process has been duplicated in laboratories.

Harold Urey and one of his students, S. L. Miller, in 1958, took a mixture of water vapor, methane, ammonia, and hydrogen—soup—and bubbled it around for a week over an electric spark.

After the week it was taken out and analyzed. It was found to contain glycine and alanine, which are called amino acids, the very stuff of life. There were also indications of three other amino acids in the soup, but in questionable form.

So out of the laboratory soup came the building blocks of life. That is what we think happened on ancient Earth.

Somehow these acids and other organic chemicals formed algaelike things. The first evidence of this in history (about 2 billion years old) was found by Elso S. Barghoorn of Harvard and Stanley A. Tyler of the University of Wisconsin in a sedimentary rock called the Gunflint chert. Later on fossil cells even older (3.4 billion years) were found, so we know life is at least that old.

There are some algae that use photosynthesis, the process of taking energy from the sun. The end product of photosynthesis is oxygen. The early forms of life were probably responsible for putting the oxygen in the atmosphere that made later life possible.

The oldest photosynthetic plants are two billion years old. (It's hard, but try to remember we are dealing with *billions* of years!)

The first animal-like creatures are called metazoa, tiny clusters of cells. They lived at the end of an era called the Proterozoic, which followed the Archean.

The rocks of this time, incidentally, record the movements of glaciers—the first of which we have evidence.

All this part of Earth's history is pretty hazy. Records get better about 600 million years ago, during an era called the Paleozoic. Like all geologic eras, the Paleozoic is divided into distinct periods, in this case seven—the Cambrian, Ordovician, Silurian, Devonian, Mississippian (Early Carboniferous), Pennsylvanian (Late Carboniferous), and the Permian. We have lots of rocks and fossils to tell us what happened.

At the time of the Cambrian period, many think the conti-

The land, the water, and the clouds. This is another *Gemini XI* photograph. It shows the southern tip of India and Ceylon, or Sri Lanka

nents were clustered together into what may be called Pangaea 1. It was generally in the Southern Hemisphere. The Gondwanaland (southern) continents were twisted together. North America and Europe were east of South America. Australia was by the equator.

There is evidence of ocean spreading even then, and by the end of the Cambrian period Pangaea 1 was splitting up. By the beginning of the next period, the Ordovician, 450 million years ago, hunks of old Pangaea were happily sailing around. Although the planet had been reasonably quiet until then, the post-Cambrian periods were fairly violent, with enormous volcanic eruptions. There were mountains being built and oceans splashed all over the place, creating new seas and leaving old ones high and dry.

The ancient Atlantic, which had opened, began shrinking at the Ordovician time. This movement of the oceans produced places where life thrived.

The equator probably ran from central Europe, through Greenland, and across North America to California and into northeast Asia. Antarctica was moving in a curve away from the equator toward the southeast.

The movement of the continents forced several plates to crash together, producing, of course, volcanoes, earthquakes, and mountains.

There was a glacier in the Sahara! That seems to be pretty good evidence that Africa was almost polar then.

The Ordovician period was followed by the Silurian, which ran to about 400 million years ago. There are some Ordovician and Silurian fossils, but they are not very helpful.

The first vertebrates came with the dawn of the Devonian period. So did a reversal in the continents' course that eventually led to Pangaea 2. The continents had changed their course and began heading back toward each other.

The oceans began to teem with fish, and the first amphibians began to crawl out onto the land. Vegetation spread to cover much of the lowlands.

The first reptiles came, and huge fern forests covered much of the land. The forests were the source of the material that makes up most of the coal we now use (which explains the name Carboniferous).

In North America these forests covered 162,000 square miles (about 414,000 km^2), and in Europe about 62,000 square miles (161,000 km^2). In these forests were the famous *Glossopteris* flora later found by Scott in the Antarctic.

At the end of the Carboniferous periods, about 270 million years ago, the Urals were formed, as were the Appalachians. As usual, these events were accompanied by great violence.

Long periods of glaciation (huge ice ages) were common during the Paleozoic era.

The last period of the Paleozoic is the Permian, the time the first mammals began to appear on Earth and the time when all the pieces slammed together forming Pangaea 2, the supercontinent Wegener described. The Atlantic ceased to exist. Europe nestled into the American coast and the late, great Tethys Sea appeared between Asia and Africa. The equator ran from the Tethys through Spain, North Africa, New England, and southern California.

The Paleozoic era teemed with life, particularly in the seas. Most of the marine animals we know evolved during that era, including the shark.

The next era is called the Mesozoic, which is divided into three periods, the Triassic, the Jurassic, and the Cretaceous.

We have a lot of Mesozoic rocks all around us. From these rocks we know the major event in the Mesozoic period, the splitting of Pangaea 2.

Huge rifts began to appear in the ground over a period of about 45 million years, starting about 150 million years ago. Mexico rode away from South America, and North America's east coast split from northwest Africa. The Atlantic began to open again. The Tethys began to shrink.

Laurasia and Gondwanaland separated and have been permanently divorced ever since.

The Pacific Ocean flooded the western United States up to where the Rocky Mountains are, and huge geosynclines were formed in that area. The rest of the land "eastward" was barren and flat until the Appalachians.

Europe was a desert. The Ural Mountains were completed.

The Jurassic period is the most famous because it was the age of the dinosaurs, the huge reptiles that ruled the world for millions of years.

Pangaea 2 continued to fragment as new rifts appeared in the ground. North and South America moved away from each other, and the space in between was filled with water.

India was moving inexorably toward Laurasia. The geosyncline in the western part of North America was producing mountains. A geosyncline along the western part of South America was being formed which would produce the Andes.

The Jurassic period ended about 130 million years ago, and the Cretaceous period began with continuing breakup of the continents and the first appearance on Earth of flowers—not a minor event.

The Laurasia-Gondwanaland formations would not be obvious on a Cretaceous globe, as the fragmentation had gone too far.

The Atlantic continued to grow from the mid-ocean ridge in the center, and the Tethys continued to be swallowed up by the crust.

The Mesozoic period was a lovely time. The glaciers had receded, in part because the continents shifted around by the equator, and the temperatures were generally mild.

The oceans were warmed, and the marine population flourished. The first flying creatures, the pterosaur (a reptile) and the earliest birds, flew during this period.

It was only toward the end of the Cretaceous that the climate began to deteriorate. Many of the more prevalent forms of life, particularly the dinosaurs, died out. No one really knows why, although climatic changes are one possibility.

The stage was set for modern times. It had taken more than 4 billion years.

The Cenozoic era, which began 60 million years ago, is the last era in the geologic timetable. We are in it now.

The era has only two periods, the Tertiary and the Quaternary. We live in the Quaternary.

Scientists divide the periods into epochs. The Tertiary period has five epochs: the Paleocene, the Eocene, the Oligocene, the

Miocene and the Pliocene. The Quaternary has but one, the Pleistocene.

In this period the Tethys disappeared and the Atlantic became the ocean we know now. The Alps, The Himalayas, and the Rockies were formed.

Greenland and Scandinavia finally broke apart. The split was dramatic. Until then the climate in what is now Norway was warm and gentle, because the waters of the ocean were warm. But with the break the cold waters from the polar regions poured in, and the climate changed drastically.

Mexico moved into its present position and was finally connected by a land bridge (Panama) to South America.

India rammed Asia, creating, of course, The Himalayas. Volcanoes roared in the western United States during the Eocene epoch, and the Pacific plate hit North America, rumpling the Rockies, during the Oligocene. Africa hit Europe and the Alps were formed during the Miocene.

And something new was stalking the Earth. It was called a primate.

The first mammal-like reptiles had been formed during the Triassic period, and a kind of mammal, called the "Mesozoic mammal," evolved late in that period and lasted into the Eocene. By the Cretaceous period evolution had produced the marsupial, whose young were born unfinished and had to be carried around in a pocket (kangaroos and opossums). The first insect eaters came about the same time. The first carnivores, meat eaters, came at about the time the Mesozoic era ended. Then came camels and elephants, whales and porpoises, rabbits and rodents. Also at about this time came the primates—about 90 million years ago.

The earliest thing we have that resembles anything from a primate is a tooth found in Montana. That is not much evidence. But there are some Paleocene fossils (75 million years old) that are much better. One fossil is of a creature that had claws and a long snout. It probably hopped around on the ground and lived in trees and looked something like a lemur.

By the Oligocene epoch creatures very much like apes evolved. We have found their remains in the Arab Republic of Egypt in a place called Faiyum.

During the Miocene epoch there lived a creature called the Ramapithecus. That was 14 or 15 million years ago. It may have been the first hominid, or humanlike creature.

When the Tertiary period ended, about 2 million years ago, the climate changed drastically for the worse. The great Ice Age descended on the land (the continents were much as we now see them), and it was during this Ice Age that the first men, Australopithecus they are called, struggled to survive.

These creatures evolved into Homo sapiens, modern man, only 30,000 years ago! In all the long history of this planet, 30,000 years is less than significant. It represents .00065 percent of the lifetime of Earth.

No one knows how long Earth will last. We know even less how long we will last on it. But it is clear that in the order of things in this world, we hardly count at all.

We do, of course, have the power to destroy life on this planet, but we probably cannot destroy the planet itself.

There are processes going on that probably will continue to go on for millions of years until the planet finally ages and goes to sleep. Or until the sun explodes, or a passing meteor shatters it, or something else mightier than man destroys it.

Earth is too big, too old, too strong for us to make much of a difference in these processes. It is true we can turn the surface into a sterile desert, but the dynamic processes that made the Earth are beyond our reach.

Earth is too much for us. We don't much matter.

In the words of the poet Yeats:

> "God stands winding His lonely horn,
> And time and the world are ever in flight;
> And love is less kind than the grey twilight,
> And hope is less clear than the dew of the morn."

Acknowledgments

There are a few people who made this book possible. They ought to get mention, if nothing else.

First and foremost is Lorna Greene Shurkin, wife and copy editor, who insisted I follow at least some of the rules of the English language. She did not have an easy time of it.

Dan Goldwater of the Franklin Institute played nursemaid, coddling me and the manuscript along.

Dr. Ken Verosub of the University of California at Davis read the manuscript and kept me honest—or at least tried. If there are any errors in this book, they exist despite his best efforts.

The pictures are here because of the kindness of Will Dooley of the U.S. Geological Survey in Reston, Virginia, and Les Gaver of the National Aeronautics and Space Administration in Washington.

Parts of the chapter on earthquakes appeared in a different form in *The Philadelphia Inquirer*, and the information was gathered while on assignment for that newspaper and national editor Steve Seplow. I am also indebted to Rob Wesson and the other scientists at the Geological Survey's Earthquake Center in Menlo Park, California, for their help, time, hospitality, and a Chinese lunch.

Thanks are also due my sons, Jonathan and Michael, who kept quiet long enough to let me finish this volume. It was out of character and they have long since reverted to their tumultuous selves.

Glossary

Abyssal Plains The flat floor of the ocean between the continental formations and the mid-ocean ridges.

Asthenosphere The plastic part of the mantle, lying below the lithosphere and above the rocky mantle. It begins about 37 miles (60 km) down and goes down to about 155 miles (250 km).

Caldera The bowl in the center of a volcanic mound.

Continental drift The theory, developed by Wegener, that the continents move about on top of the mantle, explaining the odd "fit" of shorelines apparent on a globe.

Continental rise The gentle area where the slope meets the Abyssal Plains.

Continental shelf The part of a continent under the ocean all along the shoreline.

Continental slope The towering palisades where the continental shelf dives toward the continental rise.

Convection Movement of something caused by heat.

Core The center of the Earth; generally believed to comprise the outer, liquid, section which is 1,375 miles (2,200 km) thick, and a solid core 760 miles (1,220 km) in diameter.

Crust The outer layer of the Earth—what we walk on. It is 6 to 25 miles (10–40 km) thick.

Dynamo theory The theory that explains the Earth's magnetic field by saying it is caused by the sloshing of the molten core as the Earth spins.

Epicenter The place on the surface directly above the site of an earthquake.

Fault A crack in the Earth's crust where two plates meet. The site of great geologic activity, especially earthquakes.

Geology The science dealing with the history of the Earth, particularly through the study of rocks.

Geophysics The physics of the living Earth—how it was formed, how it has moved, and the effects upon it of things like tides and weather.

Geosynclines Troughs in the Earth's crust that fill with sediment. They are eventually pushed up to form mountains.

Gondwanaland The supercontinent of the Southern Hemisphere after the breakup of Pangaea. Named by Eduard Suess after a place in India of geologic interest.

Guyot An underwater mountain with a flattened top.

Laurasia The Northern Hemisphere continent after Pangaea, named by Alex du Toit.

Lithosphere The top part of the mantle from the crust down to about 60 miles (96 km).

Magma Molten rock which, when cooled, forms igneous rock. When it flows from a volcano it is called lava.

Magnetosphere The magnetic field surrounding a planet. Earth and Jupiter are the only known planets with strong magnetospheres.

Mantle The section of the Earth that lies between the crust and the core. It constitutes most of the mass of the Earth.

Mass A quantity of matter making up one body; something like weight but more complicated.

Mid-ocean ridge The mountain ridge that runs 40,000 miles (64,000 km) around the world through the oceans. It is split by the mid-ocean rift which runs down the middle of the ridge.

Moho The Mohorovicic discontinuity, the zone believed to be where the crust meets the mantle, or 3–6 miles (5–10 km) under the ocean floor.

Nappes Folds in the crust involved in mountain-building.

Orogeny Mountain-building.

Pangaea The name Wegener gave to the land mass that comprised all the continents on the early Earth. It means "all lands."

Planet A heavenly body orbiting a sun. It can be either solid (terrestrial), like the Earth, or a thick ball of gas (Jovian), like Jupiter.

Plate Any of the huge movable segments into which the Earth's crust is divided.

Plumes Streams of heat convection currents, which push around the tectonic plates.

Pressure waves, P waves Seismic waves that move in a straight line, bulging and contracting.

Radioactivity The giving off of radiation by the decay of atomic nuclei.

Richter scale A way of measuring the energy in earthquakes, developed by Charles Richter. It is open-ended. Anything above 4.5 on the Richter scale is dangerous.

Rift A place where the continents are being split apart.

Rocky mantle The bottom part of the mantle, just above the core and below the asthenosphere. It goes down about 1,650 miles (2,650 km).

Seismic Referring to an earthquake or to the waves caused when the crust of the Earth moves, as in an earthquake.

Seismograph A machine that measures seismic waves, usually earthquakes.

Shear waves, S waves Seismic waves moving at right angles to pressure waves. *See* Pressure waves and Surface waves.

Surface waves Seismic waves that ripple at the surface, the waves that cause earthquake damage. *See* Pressure waves and Shear waves.

Tectonic Referring to changes in the Earth's crust.

Trench A steep-sided depression in the ocean floor, where the continental plates slide back into the mantle.

Uniformitarianism The idea first formulated by James Hutton that the formation of the Earth is the result of a slow, steady change throughout history.

Van Allen belts Belts of radiation surrounding the earth, trapped by the magnetosphere. Named after James Van Allen who discovered them.

Volcanism Referring to volcanoes.

Bibliography

Here is a partial list of source material used for this book. An asterisk (*) indicates a book especially recommended for young readers who would like to learn much more about some of the things I have written about. While the books are not designed for young readers, they can be most useful, and there is no rule against skipping the hard parts.

*Cameron, Ian. *Antarctica: The Last Continent.* Little, Brown & Company, 1974.

*——— *Continents Adrift: Readings from Scientific American.* W. H. Freeman & Co., 1972.

*Dinely, David. *Earth's Voyage Through Time.* Alfred A. Knopf, Inc., 1974.

——— *Exploring the Ocean World,* ed. by C. P. Idyll. The Thomas Y. Crowell Company, 1972.

Gatland, Kenneth W., and Dempster, Derek. *Worlds in Creation.* Henry Regnery Co., 1974.

Moore, Patrick. *Concise Atlas of the Universe.* Rand McNally & Company, 1974.

*——— *Watchers of the Stars.* G. P. Putnam's Sons, Inc., 1974.

Sullivan, Walter. *Continents in Motion.* McGraw-Hill Book Co., Inc., 1974.

——— *The New Treasury of Science,* ed. by Harlow Shapley, Samuel Rapport, and Helen Wright. Harper & Row, Publishers, Inc., 1965.

von Ditfurth, Hoimar. *Children of the Universe.* Atheneum Publishers, 1974.

Wertenbaker, William. *The Floor of the Sea.* Little, Brown & Company, 1974.

Index